Ozonization Manual for Water and Wastewater Treatment

Ozonization Manual for Water and Wastewater Treatment

Edited by

Dr. W. J. Masschelein
Compagnie Intercommunale Bruxelloise des Eaux, Belgium

President of the European Committee of the International Ozone Association

A Wiley–Interscience Publication

1807 1982

JOHN WILEY & SONS

Chichester · New York · Brisbane · Toronto · Singapore

Library of Congress Cataloging in Publication Data:

Main entry under title:

Ozonization manual for water and wastewater treatment.

"A Wiley–Interscience publication."
Includes index.
1. Water—Purification—Ozonization.
2. Sewage—Purification—Ozonization.
I. Masschelein, W.
TD461.0965 628.1′662 81-21986
AACR2

ISBN 0 471 10198 2

British Library Cataloguing in Publication Data:

Ozonization manual for water and wastewater treatment.

1. Water—Purification—Biological treatment
2. Sewage—Purification—Biological treatment
I. Masschelein, W. J.
628.3 TD745

ISBN 0 471 10198 2

Filmset in 'Monophoto' Times Roman
by Speedlith Photo Litho Ltd., Manchester.
Printed by Page Bros., (Norwich) Ltd.

Contents

Preface xiii

List of contributors xvi

PART 1—OZONE GENERATION

1. **Activation of oxygen in the electrical discharge and ozone formation kinetics** 3
   M. Goldman and M. Lecuiller

2. **Thermodynamic aspects of the formation of ozone and secondary products of electrical discharge** 9
   W. J. Masschelein

3. **The use of an oxygen-enriched process gas** 13
   W. J. Masschelein

4. **Practical aspects of the recycling of effluent gas into generation systems** 18
   W. J. Masschelein

5. **Alternative methods for the synthesis of ozone** 22
   W. J. Masschelein

6. **Specifications and performance control for ozone generators** . . 24
   R. Gerval

7. **Principles in supplying electrical energy to ozonators** . . . 28
   R. Mostin

8. **Generation systems** 33
   Trailigaz ozone generator technology 33
   P. Chapsal
   Degrémont's ozonators 35
   R. Louboutin
   The Litha ozonator 38
   W. J. Masschelein

Mather and Platt Capenhurst ozonator 40
R. K. W. Palmer

PART 2—FUNDAMENTAL PRINCIPLES OF THE OZONIZATION OF WATER

9. Theoretical aspects of ozone transfer into water 47
M. Roustan and J. Mallevialle

10. Stability of ozone in water 53
J. Mallevialle

11. Reaction pathways and kinetics of reactions of ozone with compounds dissolved in water 57
J. Hoigné

12. Action of ozone on humic compounds 63
D. Maier

13. Removal of bacteria and viruses by ozonization 66
J. C. Block

14. Effect of ozone on the elimination of various algae by filtration . 69
J. C. Ginocchio

15. Action of ozone on various organic compounds 73
A. Laplanche and G. Martin

16. Ozone and chlorinated organic compounds 77
M. Doré

17. Principles of repeated ozonization 81
M. Rapinat

18. Influence of ozonization on the structure of organic matter and its biodegradability within activated carbon 83
M. Bourbigot and M. Doré

19. Micellization and improvement of coagulation by ozonization . . 86
Y. Richard

20. Micellization and related effects of ozone on colloidal systems . 88
M. R. Jekel

21. Interactions of ozone with chlorine and chlorine dioxide . . . 90
J. Valenta

22. The direct action of dispersed ozonized gas bubbles 97
W. J. Masschelein

23. Perspective of combination of ozone and ultrasound 96
E. Dahi

24. Scope of joined action of ozone and ultraviolet light 103
W. J. Masschelein

25. Identified reaction products in ozonization 105
J. Mallevialle

PART 3—PRACTICAL ASPECTS OF OZONIZATION

26. Significance of the ozone to water contacting system 129
M. Bredtmann

27. Contact time of ozonization 133
J.-P. Legeron

28. Contact columns and bubble-dispersing systems 137
W. J. Masschelein

29. Materials resistant to corrosion and degradation in contact with ozone 140
F. Damez

30. Corrosion protection in ozonization 143
G. E. Kurzmann

31. Safety measures to protect the equipment 146
F. Damez

PART 4—ANALYTICAL METHODS FOR OZONE DETERMINATION

32. Methods for the determination of ozone in concentrated gas mixtures 151
D. Maier

33. Methods for the determination of ozone in diluted off-gases . . 154
L. Brener

34. Monitoring of high ozone concentration in the air. 157
J. Valenta

35. Determining the presence of ozone in work areas and in the ambient atmosphere 160
F. Damez

36. Determination of residual ozone in water 162
J. L. Hamon

37. Determination of residual ozone in water with the acid chrome violet K (ACVK) method 166
W. J. Masschelein

38. Colorimetric method for the measurement of aqueous ozone based on the decolorization of indigo derivatives 169
H. Bader and J. Hoigné

39. Spectrophotometric determination of residual hydrogen peroxide . 173
W. J. Masschelein

40. Operational analysis of residual ozone in treated water . . . 176
W. J. Masschelein

41. Principles and design criteria of automation ozonization . . 178
J. Mignot

42. Analytical instruments for ozone control 181

Degrémont residual ozone analyzer, Amperazur type . 181
J. Guiblais

In-line testing and display of the content of ozone, chlorine dioxide, chloramines and free and active potential chlorine in water . 183
A. Derreumaux and M. Lambert

The Anseros analyser for the measurement of ozone in the gas and aqueous phase 185
K. Nonnenmacher

Continuous amperometric residual ozone analysis in the Tailfer plant 187
W. J. Masschelein

The Sigrist A254 photometer for monitoring of ozonators and ozone reactions 189
W. Sigrist

PART 5—CRITICAL QUESTIONS IN OZONIZATION PRACTICE

43. Toxicity of ozone 197
F. Erb

44. Professional risks associated with ozone 200
F. Van Hoof

45. Protective equipment for emergency action 202
M. Rapinat

46. Excess ozone disposal 204
C. Coste

47. A practical type of thermal residual ozone destructor . . 209
P. Chapsal

48. Removal of the residual ozone in the air after the application of ozone 212
M. Horst

49. Methods and operating costs of ozone destruction in off-gas . . 217
K. Orgler

50. The Tailfer plant and new developments in ozone destruction techniques 220
W. J. Masschelein

51. Bacterial stabilization of ozonized water 224
H. Lamblin

52. Some aspects of bacterial aftergrowth in ozonized waters . . 227
G. Fransolet

PART 6—OZONIZATION COST

53. Global cost evaluations 233
M. Paré

54. Cost-benefit analysis of the use of an oxygen-enriched process gas . 238
W. J. Masschelein

PART 7—DEVELOPMENTS IN OZONE USE

55. Utilization of ozone in swimming pools 243
J.-P. Legeron

56. The use of ozone in wastewater treatment 248
G. Martin and M. Elmghari-Tabib

57. Some examples of ozone use in industrial wastewater treatment . 253
P. Le Tareau and J. M. Rovel

58. Possible use of ozone in papermills 257
F. A. Abadie-Maumert

59. Ozonization of bottled water 259
W. Schneider

PART 8—OZONIZATION FLOWSHEETS

60. Ozone at Nice 265
P. Decourcelle

61. The use of ozone in the treatment of water pumped from Lake Constance 267
D. Maier

62. The ozonization scheme at the Tailfer plant. 269
W. J. Masschelein

63. Ozonization unit at the plant of Crissey (Chalon-sur-Saône) . . 271
J. Mignot

64. Plant of Anconella of the city of Florence 273
J. Mignot

65. Ozonization unit at the plant of Morsang-sur-Seine . . . 276
J. Mignot

66. Ozonization unit at the plant of Roberval, Canada . . . 278
J. Mignot

67. Ozonization at the treatment plants of the 'Syndicat des communes de la Banlieue de Paris pour les Eaux' 280
M. Rapinat

68. Ozone for biological nitrification and finishing treatment of groundwater at Rouen-la-Chapelle 283
J. Caujolle-Gazet

69. Ozone at the Annet-sur-Marne treatment plant 286
A. Tiret

70. Ozonization at the Ile-au-Bourg plant at Angers 288
J. De Frescheville

71. Ozone at Rennes 289
J. De Frescheville

72. The Mauboule treatment plant at Valence 291
O. Pascal

73. The new ozonization facility at Marseille 293
J.-P. Lesbros

74. Application of ozone in the treatment of the Zürich lake water . 297
M. Schalekamp

75. Ozonization for groundwater protection at the City Waterworks of Karlsruhe 300
G. Nagel

76. The Dohne treatment plant in Mülheim-Ruhr: 'The Mülheim process' 306
M. R. Jekel

77. Ozonization and flocculation–filtration in the biological treatment of drinking water (the Essen plant) 309
A. Reicherter

78. The importance of ozonization in the purification system of the Lake Waterworks of Amsterdam 311
F. A. M. Van Soest

79. Ozonization of recycled water at Quierschied's swimming pool . 314
J.-P. Legeron

80. Ozonization at the swimming pool of Ostend 317
W. J. Masschelein

Index 319

Preface

With the growing interest in ozonization practice as applied in water treatment, the technical and scientific knowledge on this process is continuously expanding. Therefore, the time has come to provide a compendium of the information currently available internationally and used in everyday ozonization practice. The aim of this manual is to gather the key theoretical aspects of ozone, the know-how of its application and future prospects in a useful information source available to all. Accordingly, the contributors have been invited to be as accurate and concise as possible, in order to facilitate access to information useful to practical ozone users and to stimulate further development of the technique. The main objective of this handbook is to provide an immediate source of information prior to experimental investigation, and also a basis for research into defects in systems.

In a sense, synthetic ozone can be considered as the perfect imitation of chemical oxidation processes due to oxidants occurring naturally, e.g. ozone, oxygen radicals and oxygen. Natural ozone results from the UV-irradiation of the earth's atmosphere; the ground-level concentration in our areas varies between 0.08 and 8 mg per normal cubic metre. Important general disinfection properties can be related to this concentration.

On oxidizing organic molecules, ozone releases products in oxygen-containing functional groups, particularly carbonyl and carboxyl functions which support bacterial metabolic schemes. A balance must be made in the present state of the art so as to derive benefit from the enhanced biodegradability of ozonized water. This is a question of suitable technology, appropriate location of the ozonization phases in the plant flowsheet, and adequate control and operational command.

Acknowledgements

To coordinate a work by numerous co-authors is an undertaking that requires a great deal of cooperation from the contributors. Therefore, I must acknowledge and express my appreciation for their enthusiasm for this project.

Finally, I must praise my co-workers Edith Rogge and Sabine Van Laeken for having kept an eye on the timetable and my reviewing; they have lightened my work through their assistance. I must also thank André Henin who has spared no effort to improve certain illustrations.

Brussels, 1982

List of contributors

F. A. ABADIE-MAUMERT — *Norwegian Pulp and Paper Research Institute, Box 250, Vinderen, Oslo 3, Norway*

H. BADER — *Federal Institute for Water Resources and Water Pollution Control, Swiss Federal Institute of Technology, 8600 Dübendorf, Switzerland*

J. C. BLOCK — *UER d'Ecologie, Rue des Récollets 1, 57000 Metz, France*

M. BOURBIGOT — *Compagnie Générale des Eaux, 52 rue d'Anjou, 75384 Paris cédex 08, France*

M. BREDTMANN — *Gebr. Hermann—Verfahrenstechnik GmbH & Co. Grüner Weg, Postfach 300460, 5000 Köln 30 (Ehrenfeld), FRG*

L. BRENER — *Degrémont, Département Recherche Physico-Chimie, 183 avenue du 18 juin 1940, 92508 Rueil-Malmaison cédex, France*

J. CAUJOLLE-GAZET — *Compagnie Générale des Eaux, Centre Régional de Rouen, 3 rue Dufay, 76100 Rouen, France*

P. CHAPSAL — *Trailigaz, Boulevard de la Muette 29-31, 95140 Garges-les-Gonesse, France*

C. COSTE — *Degrémont, 183 avenue du 18 juin 1940, 92508 Rueil-Malmaison cédex, France*

E. DAHI — *Department of Sanitary Engineering, Technical University of Denmark, Block 115, 2800 Lyngby, Denmark*

F. Damez — *Compagnie Générale des Eaux, 52 rue d'Anjou, 75384 Paris cédex 08, France*

P. Decourcelle — *Compagnie Générale des Eaux, Boulevard de Cessole 144, 06039 Nice cédex, France*

J. De Frescheville — *Compagnie Générale des Eaux, 11 rue Kléber, 35020 Rennes cédex, France*

A. Derreumaux — *CIFEC, 10 avenue de la Porte Molitor, 75016 Paris, France*

M. Doré — *UER Sciences fondamentales et appliquées, 40 avenue du Recteur Pineau, 86022 Poitiers cédex, France*

M. Elmghari-Tabib — *CEEA–ENSC Rennes, avenue du Général Leclerc, 35000 Rennes-Beaulieu, France*

F. Erb — *Faculté de Pharmacie de Lille, Universite de Lille II, Place de la Victoire 6, 59130 Lambersart, France*

G. Fransolet — *Compagnie Intercommunale Bruxelloise des Eaux, 764 chaussée de Waterloo, 1180 Bruxelles, Belgium*

R. Gerval — *Syndicat des Communes de la Banlieue de Paris pour les Eaux, La Tour de Lyon, 185 rue de Bercy, 75012 Paris, France*

J. C. Ginocchio — *Water and Wastewater Treatment Department, Sulzer Brothers Limited, 8401 Winterthur, Switzerland*

M. Goldman — *Ecole Supérieure d'Electricité, Plateau du Moulon, 91190 Gif-sur-Yvette, France*

J. Guiblais — *Service Appareils et Asservissements, Degrémont, BP 46, 92151 Suresnes cédex, France*

J. L. Hamon — *Compagnie Générale des Eaux, 52 rue d'Anjou, 75384 Paris cédex 08, France*

J. Hoigné — *Federal Institute for Water Resources and Water Pollution Control, Swiss Federal Institute of Technology, 8600 Dübendorf, Switzerland*

M. HORST — *Bismarckstrasse 4, 6905 Schriesheim, FRG*

M. R. JEKEL — *DVGW–Forschungsstelle am Engler–Bunte Institut, Postfach 6380, 7500 Karlsruhe 1, FRG*

G. E. KURZMANN — *O. W. T. Mathias-Hess Strasse 40, Postfach 1414, 6909 Wallorf, FRG*

M. LAMBERT — *CIFEC, 10 avenue de la Porte Molitor, 75016 Paris, France*

H. LAMBLIN — *Compagnie Generale des Eaux, 52 rue d'Anjou, 75384 Paris cédex 09, France*

A. LAPLANCHE — *CEEA–ENSC Rennes, avenue du General Leclerc, 35000 Rennes-Beaulieu, France*

M. LECUILLER — *Ecole Supérieure d'Electricité, Plateau du Moulon, 91190 Gif-sur-Yvette, France*

J.-P. LEGERON — *Trailigaz, 29–31 boulevard de la Muette, 95140 Garges-les-Gonesse, France*

J.-P. LESBROS — *Société des eaux de Marseille, 55 boulevard des Acieries, 13010 Marseille, France*

P. LE TAREAU — *Degrémont, 183 avenue du 18 juin 1940, 92508 Rueil-Malmaison cédex, France*

R. LOUBOUTIN — *Degrémont, 183 avenue du 18 juin 1940, 92508 Rueil-Malmaison cédex, France*

D. MAIER — *Betriebshabil- und Forschungslabor der Bodensee-Wasserversorgung, 777 Überlingen-Süssenmühle, FRG*

J. MALLEVIALLE — *SLEE, 38 rue du President Wilson, 78230 Le Pecq, France*

G. MARTIN — *CEEA–ENSC Rennes, Avenue du General Leclerc, 35000 Rennes-Beaulieu, France*

W. J. MASSCHELEIN — *Compagnie Intercommunale Bruxelloise des Eaux, Chaussée de Waterloo 764, 1180 Bruxelles, Belgium*

J. M. MIGNOT — *Degrémont, 183 avenue du 18 juin 1940, 92508 Rueil-Malmaison cédex, France*

R. MOSTIN	*Compagnie Intercommunale Bruxelloise des Eaux, 70 rue aux Laines, 1000 Bruxelles, Belgium*
G. NAGEL	*Stadtwerke Karlsruhe, Daxlarder Strasse 72, 7500 Karlsruhe 1, FRG*
K. NONNENMACHER	*Anseros, Hauptstrasse 46, 7031 Gärtringen, FRG*
K. ORGLER	*Trailigaz, Boulevard de la Muette 29–31, 95140 Garges-les-Gonesse, France*
R. K. W. PALMER	*Mather & Platt, Victoria Road, Burgess Hill, Sussex RH15 9LH, UK*
M. PARÉ	*Degrémont, 183 avenue du 18 juin 1940, 92508 Rueil-Malmaison cédex, France*
O. PASCAL	*Compagnie Générale des Eaux, 75 rue Cuvier, 69455 Lyon cédex 3, France*
M. RAPINAT	*Compagnie Générale des Eaux, 52 rue d'Anjou, 75384 Paris cédex 08, France*
A. REICHERTER	*Stadtwerke Essen AG, Rüttenscheider Strasse 27–37, D-4300 Essen 1, FRG*
Y. RICHARD	*Degrémont, 183 avenue du 18 juin 1940, 92508 Rueil-Malmaison cédex, France*
M. ROUSTAN	*INSA Chimie, 31077 Toulouse cédex, France*
J. M. ROVEL	*Degrémont, 183 avenue du 18 juin 1940, 92508 Rueil-Malmaison cédex, France*
M. SCHALEKAMP	*Wasserversorgung Zürich, Hardhof 9, Zürich, Switzerland*
W. SCHNEIDER	*Fresenius Institut, Hochenstrasse 2, 6209 Heidenrod 72, FRG*
W. SIGRIST	*Sigrist Photometer AG, Zweierstrasse 129, 8036 Zürich, Switzerland*
A. TIRET	*Société Française de Distribution d'Eau, 5 place de l'Arche Guedon, 77200 Torcy, France*
J. VALENTA	*Wasserversorgung Zürich, Hardhof 9, 8093 Zürich, Switzerland*

F. VAN HOOF — *Antwerpse Waterwerken NV, Mechelsesteenweg 64, 2000 Antwerpen, Belgium*

F. A. M. VAN SOEST — *Gemeente Waterleidingen Amsterdam, Condensatorweg 54, Postbus 8169, 1005 AD Amsterdam, The Netherlands*

1. OZONE GENERATION

CHAPTER 1

Activation of oxygen in the electrical discharge and ozone formation kinetics

M. GOLDMAN and M. LEGUILLER

Ecole Supérieure d'Electricité

INTRODUCTION

The global equation representing the formation of ozone is as follows:

$$3\,O_2 \rightarrow 2\,O_3 \quad \Delta H = 68\,\text{kcal} \tag{1}$$

this energy of 68 kcal supposes that the molecules O_2 and O_3 are to be found in their fundamental state (respectively $3\,\Sigma\,\bar{g}$ and ^1A).

We may deduce that the maximum energetic yield will be of 1220 g/kWh. The actual industrial systems enable to obtain a maximum energetic output of approximately 5% (~60 g/kWh), i.e. quite far from the theoretical maximum.

The energy of 34 kcal necessary for the formation of one mole ozone cannot be brought in thermically; indeed, this would mean temperatures of gas which would deplace the equilibrium of the reaction (1) to the left. The energy must be furnished by other means such as by atoms of oxygen, the excited states of the molecule, or by the ions (negative or positive). Here we shall comment on the creation of these types of activation of the oxygen. Overheating of the gas during the procedures which favours the formation of the latter must be as weak as possible. It is in 'cold' plasma such as formed by the corona-discharge that this condition is fulfilled best. Even with such plasma, the rise in temperature of the gas contributes to the limitation of the yield.

NATURE OF THE INTERMEDIATE KINDS (activated oxygen)

The mechanism the most referred to for the formation of ozone in the electrical discharges is the following:

$$O + O_2 + M \xrightarrow{k_1} O_3 + M$$

The function of the third element M is to absorb (under kinetic or potential form) the excess energy of the reaction. The constant velocity k_1 depends on the nature of the third element[1] and on the state of excitation of the species involved, as written in the left-hand side of the equation.

If in most cases this reaction is predominant other pathways implicating several other intermediate species may not be neglected *a priori.*

Table 1 indicates the principal ones in which intervene the negative ions O_n^-, the positive ions O_n^+ and the excited molecular states O_2^*.

Table 1

Reactions		Velocity constant*
(1) $O^- + O_2 + O_2 \rightarrow$	$O_3 + O_2^-$	$32\,10^{12}$
(2) $O_3^- + O \rightarrow$	$O_3 + O^-$	$6\,10^{8}$
(3) $O_3^- + O_2^+ \rightarrow$	$O_2 + O_3$	$12\,10^{14}$
(4) $O_3^- + O^+ \rightarrow$	$O + O_3$	$12\,10^{14}$
(5) $O_2^- + O^+ + O_2 \rightarrow$	$O_3 + O_2$	$72\,10^{16}$
(6) $O_3^+ + O_2^- \rightarrow$	$O_3 + O_2$	$12\,10^{14}$
(7) $O_4^+ + O \rightarrow$	$O_2^+ + O_3$	$3\,10^{11}$
(8) $O_2^* + O_2 \rightarrow$	$O_3 + O$	Depends on the excitation
(9) $O_2^* + O_2^* + O_2 \rightarrow$	$2O_3$	level of O_2^*

* The kinetic constants are indicated as values at ambient temperature within an electrical yield and expressed in l/mole . s in the case of diparticle reaction and in $l^2/mole^2$. s for tri-particle reactions.

Remark: O_2^* represents a molecule O_2 in an excited electronic state. The energy of this state must be higher than 34 kcal/mole above that of the fundamental state so as to enable the reaction (9) and more than 100 kcal/mole for reaction (8).

ORIGIN OF THE INTERMEDIATE SPECIES

These species result from a primary process of the type:

$$e + O_2 \rightarrow \text{products}$$

Table 2 gives the list of the different types of processes and refers to the literature for further details. All the procedures quoted are not equally probable. Of course, their effective sizes do depend on the energy of the electrons implied. To be able to evaluate the relative contributions of these different procedures we would have to know each time the function of the electronic distribution in the discharge, which is far from being possible now for several types of discharges. This is the case of the corona discharge for instance. However, in certain types of discharges with homogeneous geometry, computation of electronic distribution function is possible. This is how Masek *et al.*[2] firstly and Fournier *et al.*[3] secondly, calculated the velocity constants of the procedures in function of the reduced electrical field E/N.

Table 2

Reactions	Threshold energy (kcal/mole)
(1) $e + O_2 + O_2 \rightarrow O_2^- + O_2$	0
(2) $e + O_2 \rightarrow O + O^-$	80
(3) $e + O_2 \rightarrow O + O + e$	117
(4) $e + O_2 \rightarrow O_2^+ + 2e$	300
(5) $e + O_2 \rightarrow O^- + O^+ + e$	400
(6) $e + O_2 \rightarrow O_2^* + e$	Depends on the excited state reached

OZONE DESTRUCTION MECHANISMS LIMITING ITS FORMATION

Case of pure oxygen

Thermic decomposition:

$$O_3 + O_2 \rightarrow O_2 + O_2 + O$$

This reaction, the reverse of that producing ozone, has as velocity constant of:

$$k = 2.02\,10^{-12} \exp^{-(24000/RT)} \text{l/mole.s}$$

ozone can also be destroyed by the very atoms of oxygen: $O_3 + O \rightarrow 2\,O_2$, this reaction has a velocity constant of:

$$k = 3{,}37\,10^{10} \exp^{-(5700/RT)} \text{l/mole.s.}$$

The fact that these reactions strongly depend on the temperature proves the necessity to reduce as much as possible the temperature of the gas.

The collisions with charged species may also destroy the ozone molecule; Table 3 gives the list of most of them.

Role of the 'impurities'

It is quite out of question to consider the role of all impurities liable to exist in the gas; if only taking into account the usual components of the air (N_2, NO_2, CO_2, H_2O and O_2), it would already be necessary to use several hundreds of elementary reactions.[4–6] In this paper we shall only examine the very important role of the water vapour.

Table 3

Reactions	Velocity constant (l/mole . s)
(1) $e + O_3 \rightarrow O_2^- + O$	0.23
(2) $e + O_3 \rightarrow O^- + O_2$	$1.8 \ 10^9$
(3) $e + O_3 \rightarrow O_3^- + h\nu$	$6 \ 10^3$
(4) $e + O_3 \rightarrow e + O + O_2$	$1.2 \ 10^{10}$
(5) $O^- + O_3 \rightarrow O + O_3^-$	$3 \ 10^{11}$
(6) $O^- + O_3 \rightarrow O_2 + O_2 + e$	$1.8 \ 10^{11}$
(7) $O^- + O_3 \rightarrow O_2^- + O_2$	$6 \ 10^9$
(8) $O_2^- + O_3 \rightarrow O_3^- + O_2$	$2.1 \ 10^{11}$

It is generally observed that the presence of steam in the gas submitted to discharge brings about a decrease in the ozone production. The most probable interpretation would be the following chain-reaction mechanism:

$$OH^v + O_3 \rightarrow HO_2 + O_2$$

$$HO_2 + O_3 \rightarrow OH^v + 2\,O_2$$

etc.

OH^v stands for the OH radical excited vibrationally; which can be obtained through the discharge itself

$$e + H_2O \rightarrow e + H + OH^v$$

or by the intermediate of excited atoms of oxygen in the state ^1D

$$O^1D + H_2O \rightarrow OH + OH^v$$

For some types of discharge (negative crown discharge between peak and plan) the decrease in the ozone yield in presence of steam is not observed;[7] this phenomenon still remains without interpretation (Figure 1).

Beyond this diminution of the yield, the presence of steam in an electrical discharge in air leads to the undesirable formation of nitric acid resulting from the discharge. This last statement in itself is sufficient to justify a thorough drying of the gas which is to be effluvated.

MODELLING

In the discharges producing cold plasma capable of inducing the formation of ozone, the thermodynamic equilibrium is far from being reached and it is not possible to foresee theoretically the production of ozone on the basis of the thermodynamics in the fundamental state.

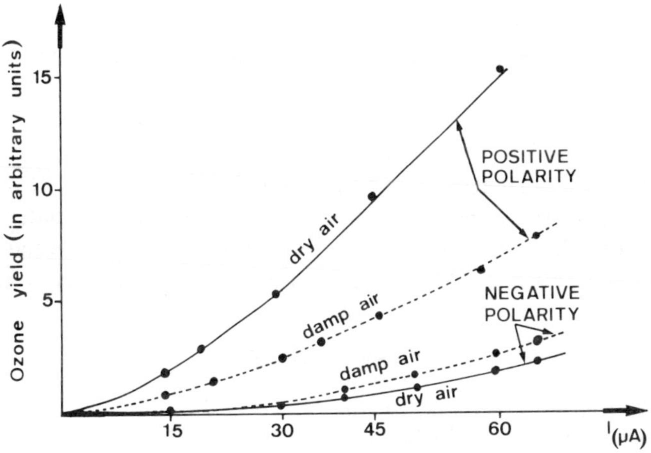

Figure 1. Effect of water vapour on ozone producing electrical discharge

Seeing a great number of data are available concerning efficient sections or the velocity constants of the elementary steps, it is feasible to start with kinetic modelling.

Several attempts have already been made in this direction. A model using only certain reactions in pure oxygen[8,9] has shown that the maximum output strongly depends on the efficient sections of the reactions producing the atoms of oxygen. The present still incomplete data on these efficient sections, lets us suppose a maximum energetic production of approximately 400 g/kWh; this model also stresses the influence of the temperature according to the experimental results. In ambient air, a model using approximately 100 reactions showed the important role of the water vapour in the formation of intermediate species in a negative crown discharge.[6]

Finally, the formulation of models of a homogeneous discharge (discharge controlled by electron-beam)[3,10] gives results quite according to the experience, in particular for the influence of the reduced electric field.

CONCLUSION

It seems that the principal reaction for the formation of ozone from oxygen in an electrical discharge is:

$$O + O_2 + M \rightarrow O_3 + M$$

Only the zone of the discharge where the electrical field is high, is liable to be active, that is in the luminous zone that spreads in the inter-electrodes space called streamer.

It is advised to consider precisely the parameters of the discharge having an influence on the output.

The fundamental macroscopic parameter is the discharge current.[7] One of the approaches consists of a decrease in minimum voltage necessary to obtain a given current, e.g. by diminishing the distance between the electrodes.

It is possible to show[11] that the discharge of the crown type makes use of two components of the current: the one, in phase with the voltage, called synchronous component, and the other, an impulsed component, associated with the streamers when the yield in ozone is at maximum. This explains the interesting yields obtained by some authors using an impulsed applied tension.[12–14]

REFERENCES

1. J. V. Michael, *J. Chem. Phys.*, **54** (10), 4455–9 (1971)
2. K. Masek, L. Laska and T. Ruzicka, *Czech. J. Phys.*, **B28**, 1321–34 (1978)
3. G. Fournier, J. Bonnet, J. Fort, D. Pigache, and M. Lecuiller, *4th Int. Symp. on Plasma Chemistry, Proceedings*, vol. 2, pp. 742–7, Zurich (1979)
4. F. E. Niles, *J. Chem. Phys.*, **52** (1), 408–24 (1970)
5. M. N. Hirsh, *Final Report on Contract no. DAHCO4-74-G-0045*. Prepared for U.S. Army Research Office (1975)
6. F. Bastien, R. Haug, and M. Lecuiller, *J. Chimie Physique*, **72** (1), 105–12 (1975)
7. F. Cribier, A. Goldman, and M. Lecuiller, IEEE *Int. Symp. On Electrical Insulation, Proceedings*, pp. 282–5, Montreal (1976)
8. R. C. Davidson, *IEEE–IAS Conf. on Electrostatics*, Pub. no. 78 CH 1346–61 A. Toronto (1978)
9. R. C. Davidson and M. Lecuiller, *4th Int. Symp. on Plasma Chemistry, Proceedings*, vol. 2, pp. 723–7. Zurich (1979)
10. G. Fournier, D. Pigache, and M. Lecuiller, *J. de Physique*, in press.
11. E. Crochet, M. Goldman, and R. Haug, *2nd Int. Symp. on Gaseous Dielectrics*. Knoxville, USA (1980)
12. L. M. L. F. Hosselet, *Electrochim. Acta*, **18**, 1033–41 (1973)
13. F. Lowther (Union Carbide), *Brevet français*, no. 7607387 (1969)
14. B. Eliason and U. Kogelschatz, *4th Int. Symp. on Plasma Chemistry, Proceedings*, vol. 2, pp. 729–34. Zurich (1979)

CHAPTER 2

Thermodynamic aspects of the formation of ozone and secondary products of electrical discharge

W. J. MASSCHELEIN

Compagnie Intercommunale Bruxelloise des Eaux

Thermodynamically, if we take the reaction of formation of ozone globally into consideration, we obtain:

$$285\,\text{kJ/mole} + 3\,O_2 \rightarrow 2\,O_3$$

In other words, the internal energy of ozone is approximately 143 kJ/mole higher than that of oxygen in the fundamental state. Thus, theoretically, 2960 Joule, or 0.82 Wh, would be sufficient to produce one gramme of ozone. Nevertheless, in practice, for the actual high-performance ozonators, the quantity of energy necessary lies between 15 and 20 Wh/g ozone produced from air. The thermodynamic output of the ozone production from air only reaches 5 % with difficulty and this at the nominal production of a given equipment.

The fundamental processes that intervene in the formation of ozone implicate the intermediate presence of activated oxygen such as free radicals at the state ^3P or ^1D and also ionized oxygen species. The different stages are illustrated in the curve of the oxygen potential given in Figure 1. The principal formation reactions of ozone from oxygen dissociated in the electronic impact are given as follows:

$O_2 + e \rightarrow O\ (^3P) + O\ (^1D)$	-7.083 eV
$O\ (^3P) + O_2 + M \rightarrow O_3 + M$	$+1.084$ eV
$O\ (^1D) + O_2 \rightarrow O_3$	$+3.051$ eV
	-2.948 eV

For the oxygen as free radicals in the state (^3P) a triparticulate choc involving a molecule of oxygen is necessary. On the other hand, from activated oxygen ^1D the

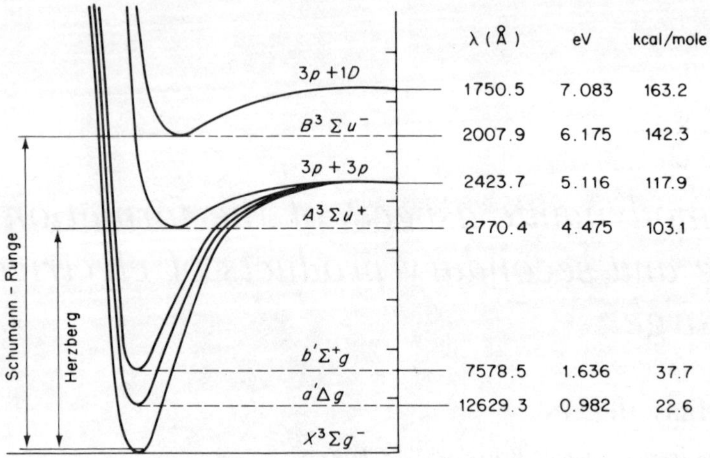

Figure 1. Potential curve of oxygen

direct formation is possible through a biparticulate choc with a molecule of oxygen.

Furthermore, under the influence of the impact of electrons a part of the ozone formed may be decomposed by ionization reactions:

$$e + O_3 \rightarrow O_2 + O^- \qquad (+\ 42\,\text{kJ/mole})$$
$$O^- + O_3 \rightarrow O_2 + O^- \qquad (+343\,\text{kJ/mole})$$
$$O(^3P) \text{ or } (^1D) + O_3 \rightarrow 2\,O_2 \qquad (+390\,\text{kJ/mole})$$

Whereas, in the whole of the reactions for formation–destruction of ozone, the presence of nitrogen interferes with the reactions. These latter lead to the formation of nitrogen oxides of which the most important component is nitrogen pentoxide N_2O_5. The formation under a high voltage ($>10{,}000$ V) is all the more important when the electrical discharge is illuminant or a spark discharge. This situation is favoured, among others, by the presence of water vapour, dust particles, organic compounds, etc.

In normal conditions, and with dry air (dew point $< -50°C$), for the ozonators at 50 Hz and applying a voltage from 12 to 20 kV, the formation of N_2O_5 reaches 1 mole (110 g) per hundred moles of ozone (4.8 kg).

In presence of traces of water, the hydrolysis of the nitrogen pentoxide gives rise to the formation of nitric acid: $N_2O_5 + H_2O = 2\,HNO_3$. In normal practice, the formation of nitric acid occurs on the basis of 3 to 5 g per kg of ozone formed.

By thermic decomposition, the nitrogen pentoxide can form the dioxide liquefying at 21°C and less:

$$N_2O_5 \rightarrow NO_2 + 1/2\,O_2$$

On the other hand, the radiations of wavelengths smaller than 3.660 Å induce reactions interfering directly with the output formation—decomposition of ozone:

$$h\nu + NO_2 \rightarrow NO + O^0$$

$$O^0 + NO_2 \rightarrow O_2 + NO$$

$$NO + O_3 \rightarrow NO_2 + O_2$$

$$NO_2 + O_3 \rightarrow (NO_3) + O_2$$

$$(NO_3) + NO_2 \rightarrow N_2O_5$$

The set of the chain reactions involving nitrogen oxides could therefore be far more important as factors limiting the formation of ozone than the concentration in N_2O_5 observed could lead us to suppose. The practical influence is all the more important as the air submitted to the discharge is less pretreated so as to dry it and remove dust particles.

If the thermodynamic conditions of the transfers between nitrogen oxides are well established, their practical formation mainly depends on the kinetic conditions still subject to further study. The problem is of a certain importance in the field of the evaluation of the toxicity of ozone in air as the synergic effects during the breathing of nitrogen oxides and ozone have been pointed out.

In the presence of vapour in the discharge zone, the output of ozone formation decreases. The following reactions principally intervene:

$$OH^0 + O_3 \rightarrow HO_2^0 + O_2 \quad (k = 1.610^{-12} \exp(-1000/T)$$

$$OH^0 + O(^1D) \rightarrow HO_2^0$$

$$HO_2^0 + O_3 \rightarrow HO + 2O_2 \quad (k = 10^{-13} \exp(-1250/T)$$

$$NO_2 + OH^0 + M \rightarrow HNO_3 + M$$

Other gases, even only at traces, do generally determine a loss of yield in formation of ozone: H_2, F_2, Cl_2.[1] Only CO and perhaps nitrogen at weak concentrations in oxygen can slightly increase ($<10\%$ additional of formation) the quantity formed in given electrical conditions. CO_2 can have a favourable effect on the formation yield of ozone.[1] The present trend is to use ternary mixtures N_2–O_2–CO_2 to uprate the production capacity of a given equïpment.

To conclude: the ozonator with laminar discharge is a reactor at the stage of equilibrium of which the thermodynamic output remains weak, i.e. below 5%. The most important part of the energy engaged is dissipated in the form of heat. The effluvated gas has a composition corresponding to a 'steady state' resulting from the kinetic evolution of the reactions occurring in the effluvated gas. In this, the dominant parameters are the reduced electrical field and current intensity, gas composition and residence time in the discharge zone.[1]

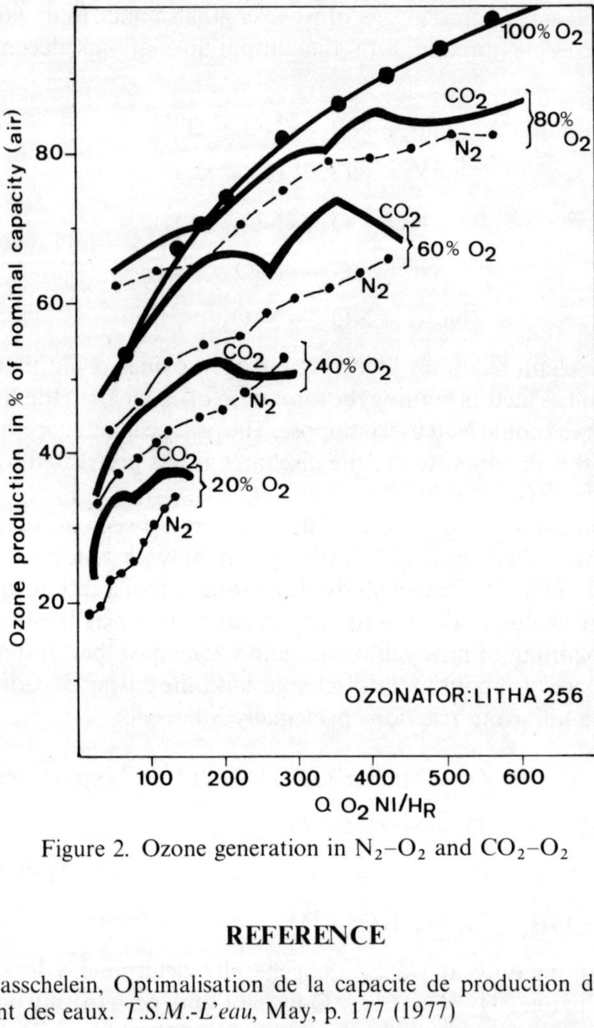

Figure 2. Ozone generation in N_2–O_2 and CO_2–O_2

REFERENCE

1. W. J. Masschelein, Optimalisation de la capacite de production d'ozone dans le traitement des eaux. *T.S.M.-L'eau*, May, p. 177 (1977)

CHAPTER 3

The use of an oxygen-enriched process gas

W. J. MASSCHELEIN

Compagnie Intercommunale Bruxelloise des Eaux

The most practical method used for the synthesis of ozone in water treatment plants remains the corona-type electrical discharge in a gas containing oxygen.

In this method several mechanisms of formation intervene as the ozonator is to be considered as a chemical reactor at equilibrium conditions resulting from processes developing in opposite ways and this as a global function of a series of parameters imposed on the system, i.e. the density of the current, the pressure, the thickness and the nature of the dielectrics, the qualitative composition of the discharged gas, etc.

Among these, the oxygen content is of the foremost importance and it is often considered that the use of pure oxygen in relation to air enables one to 'double' the quantity of ozone obtained.

These considerations generally concern the concentration of ozone in the process gas rather than the quantity of ozone produced per unit of time under constant working conditions.

For the most widespread technology, such as tube-ozonators fed with low frequency current (50 or 60 Hz) and in the nominal working conditions, the variation of ozone production in function of the oxygen content of the discharged gas is to be noted as indicated in Figure 1. For the comparison between several technologies to be valuable, we must express the results in a normalized way. In this case the recommended process is the following: for a given $\%\, O_2$ in volume one obtains $\%\ N_2 = 100 - \%\, O_2$, neglecting the traces of other gases. In a mixture of oxygen-enriched air, nitrogen is only brought in by the air. Whence:

$$\text{Q air} = \frac{100\,\text{N}\,\%}{79}$$

The volumic part of oxygen introduced by the air is equal to:

$$\frac{21\,\text{Q air}}{100} = \frac{21\ (\%\,\text{N}_2)}{79}$$

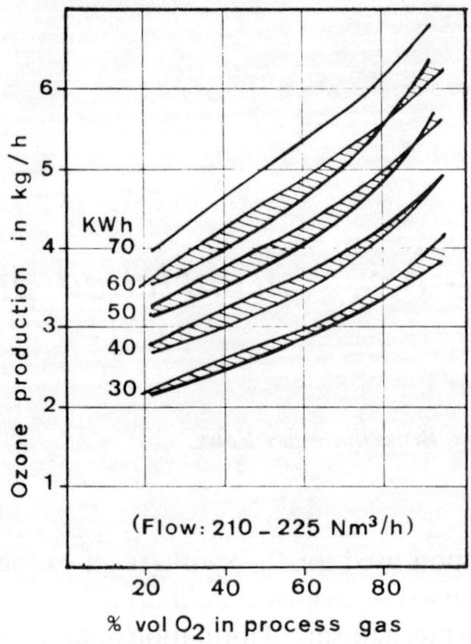

Figure 1. Ozone production per hour at constant (nominal) flow

The volume of pure oxygen brought in is equal to:

$$(100 - N_2\,\%) - \frac{21\,N_2\,\%}{79} = \frac{7900 - 100\,(N_2\,\%)}{79}$$

The relative input of oxygen in relation to the air becomes:

$$\frac{Q\,O_2\ (\text{input}) \times 100}{Q\ \text{air}} = \frac{100\,(79 - N_2\,\%)}{N_2\,\%}$$

This ratio can directly be correlated to the relative increase in production:

$$\frac{100\,DP}{P} = \frac{100\ (\text{production observed} - \text{production with air})}{\text{production with air}}$$

Figure 2 gives a few examples of this normalized method of control of the performances.

The constructive parameters intervening in these differences were studied more in detail in previous literature.

The practical use of oxygen or O_2-enriched air as discharge gas involves some specific technical aspects.

At the scale of the drinking water treatment plant applying the post-ozonization technique, a storage of liquefied oxygen is usually more practical.

As an alternative, the air may be enriched itself by the technique adsorption–desorption by pressure-swing. However the major drawback of this method is that it causes an objectionable noise in the environment.

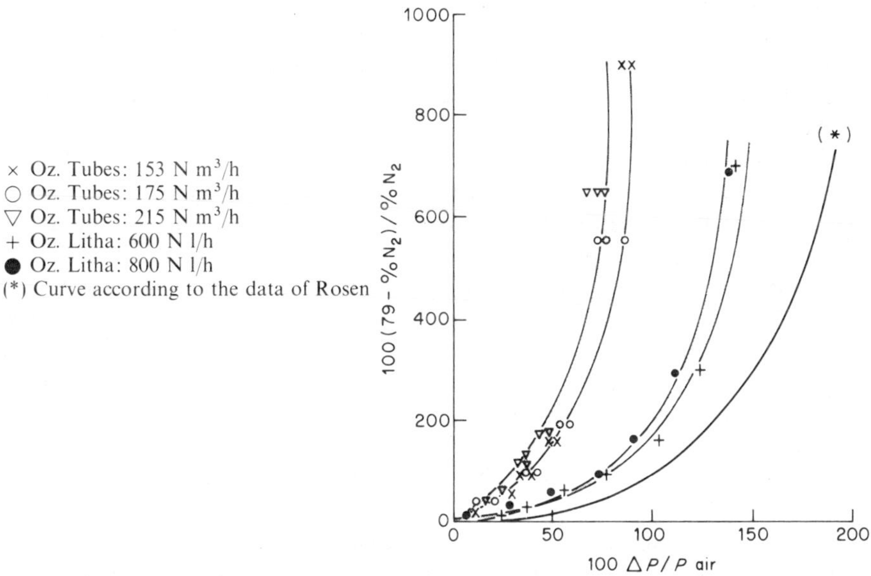

Figure 2. Increase in the production of ozone through simpsonizing of the air with oxygen

Figure 3. Stockage and expansion of liquid oxygen at the Tailfer plant (CIBE)

The liquefied oxygen stored under pressure (at approximately 15 bar) is evaporated through atmospheric evaporators and transferred through a gas circuit at the pressure of 1 to 3 bar. At the dosing posts the gas is expanded to a value slightly higher (e.g. 0.7 bar) than that of the primary air circuit to be effluvated.

The introduction takes place in the circuit downstream of the drying towers and upstream of the ozonators in electrical discharge. The oxygen content is determined with an appropriate analyser. The measurement of the paramagnetism is therefore the most suitable.

In parallel with the increase in the capacity of ozone production, the enrichment in oxygen of the discharged gas determines an increase in dissolved oxygen of the ozonized water and secondarily a partial degasification of the dissolved nitrogen in the water submitted to ozonization. The global schedule fits into the pattern of the dissolution and degasification equilibrium as a function of the total pressure.

This factor intervenes among others in the case when recycling of the process gas brought in contact with water is considered or used.

The ecoomical aspects of the use of oxygen in the ozonization process has provoked several discussions. Briefly, expressed in weight of ozone obtained, the use of oxygen determines a higher cost price of the ozone obtained with a given technology of which we only consider the energetic cost and the cost in oxygen. However, if we take into account the whole of the costs, the maintenance and the paying off of the installation included, the usage of oxygen according to the enriching technique has proved to be a competitive alternative, even favourable, mainly if variable ozone needs demand a flexible production.

Typical data issuing from the Tailfer plant are as follows: the maximum capacity able to meet maximal needs can be expressed as 80/100 nominal production capacity × 1.5 (1.75) which is the possible increase in production capacity obtainable by using a process gas enriched with oxygen. Whence a nominal production capacity of 1.2 to 1.4 times the minimal needs becomes sufficient to meet a maximal ozone demand of 200 % of the minimum demand. This is equal to a scale economy of at least 1.3 to 2 or 35 % of the installed production capacity (which can reach 50 % reduction in some cases).

Table 1 Ratio of capital investments at the scale of the Tailfer plant in 1976

	MF
Nominal production capacity	90
Uprating oxygen by 75 % of nominal	10
Economy of investments (0.75 × 90 − 10)	57.5
Reactualization (1980)	84
Or an annuity of	8

(MF = million Belgian francs; 24,000 US $ 1981).

Reported on a yearly basis of 75 % of the nominal capacity of the plant, this is equal to a difference of 0.11 BF/m^3 or 3 to 4 times the operational cost of oxygen, which amounts (1978 to 1980) at 0.02–0.03 BF/m^3 treated water (510^{-4} to 710^{-4} US \$/m^3).

Therefore, besides its operational advantages, the direct use in ozone generation of an oxygen-enriched process gas is confirmed to be economically feasible and meet variable ozone demands, peak water consumption periods or other necessary variable operation rates of the plants.

FURTHER READING

M. Simon and H. Scheidtmann, *G.W.F.*, **32,** 877 (1968)
W. J. Masschelein, G. Fransolet, J. Genot, and R. Goossens, *T.S.M.-L'eau*, **76,** 385 (1976)
W. J. Masschelein, *T.S.M.-L'eau*, **77,** 177 (1977)
W. J. Masschelein, *Congress 10 A*, Berlin (1981)
H. Rosen, *Water and Sewage Works*, **119,** 114 (1972)

CHAPTER 4

Practical aspects of the recycling of effluent gas into generation systems

W. J. MASSCHELEIN
Compagnie Intercommunale Bruxelloise des Eaux

The potential advantages associated with the use of oxygen or air charged in oxygen as effluvated gas in the synthesis of ozone can be summarized as follows:

(a) the possibility of obtaining a higher ozone concentration in the gas allowing an easier dissolution and faster reaction velocities, especially in mass-transfer conditioned kinetics;

(b) the possibility of increasing momentarily the ozone production of a given plant and thus a great flexibility in the production capacity with less investments.

The potential drawbacks for the use of discharged gas enriched in oxygen appear first of all to be of economic order considering the cost of the oxygen necessary. Generally, the recycling of a gas after the contact of ozonized gas with water is considered favourable. At this stage appears the necessity to study more thoroughly the aspects in relation to the equilibrium solubility of the oxygen, nitrogen, and even CO_2 in water in function of the temperature, the pressure, the relative proportions of the different gases and, accessorily the salinity of the water.

As a preliminary for the ozonation as an incorporated process in the treatment schemes, the water is generally in equilibrium of dissolution of gases at the atmospheric pressure defined by the air containing 21 % oxygen and 79 % nitrogen.

If we put in contact with an adequate dispersion process, a gas containing more oxygen, the consequence is an increased dissolution of this latter but also a degasification of the nitrogen previously dissolved in water contacted with the process gas. So, the gas exhausted after dissolution of ozone has a lowered oxygen content and an increased proportion of nitrogen compared to the inflowing gas. The recycling of the vented gas implies thus the bleed-off of a certain volume to evacuate the excess nitrogen of the recycled gas and thus also an extra input of oxygen to set up again the composition of the gas to be effluvated.

The practical and economic consequences relating to this system of bleed-off *vs* transfer depend on the conditions of pressure, temperature and the ratio gas to liquid in the contacting system. Several approaches are described in the literature,[1–5] taking into account the practical processes applied.

In contactors at the atmospheric pressure or at pressures of the same order, the theoretical balance is established accormng to the solubility of the gas engaged and to the law of Henry. However, the degasification of nitrogen does not reach the yield of 100 % in relation to the theoretical value. The rate of degasification obtained practically depends on the whole system of putting into contact of the gas with treated water and must be determined for each particular case. An example is detailed previously.[1]

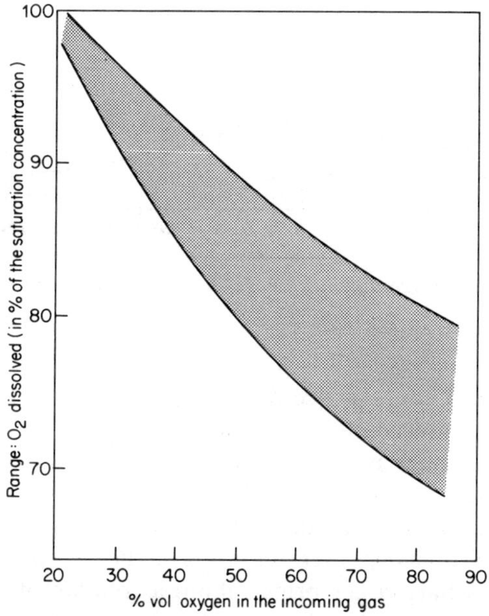

Figure 1. Average percentage saturation in O_2 as a function of the O_2 content of the ozonization process gas

The concentration in dissolved oxygen at the saturation of the water increases linearly in function of the pressure. This effect limits itself in practice to the finding of the pressure corresponding to the oxygen transfer necessary in the treatment of raw waters or of sewage water for instance. Correlatively, the gaseous nitrogen contained in the equilibrating phase diminishes in relation to the oxygen as the concentration in dissolved oxygen increases. This relation is a type of inverse function.

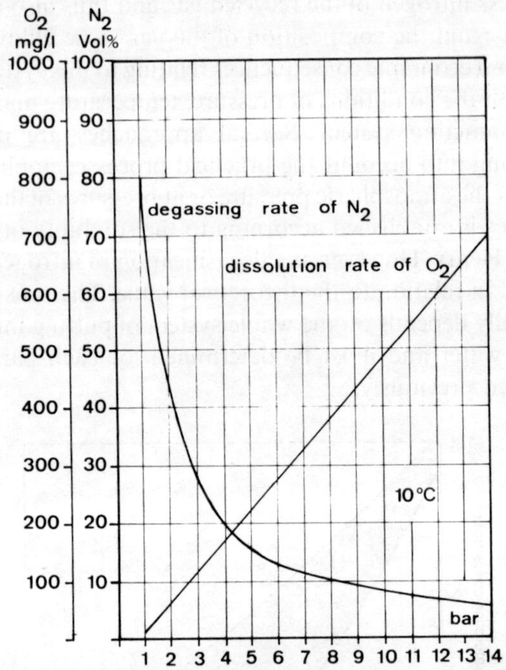

Figure 2. Solution equilibrium O_2/N_2 for water saturated in N_2 at 10°C

This can have as a consequence that the recycling of a gas to be effluvated necessitates less or no bleed-off to evacuate the nitrogen of the recycled gas in a dissolution system under pressure. The practical limit of the system, which quantitative relations are detailed in the literature[4,6] consists in the content in dissolved oxygen normally determined by excess of need, for instance 300 g/m^3 at 8 bar pressure applied on the water. Therefore, in the high-pressure technique the recycling of the discharged gas brings about a solution of the ozone and the equilibration of the water in the relation to the oxygen–nitrogen of a partial water flow that is afterwards diluted again in the whole of the water flow to be treated. The practical parameters for the conception of the whole and for operating are:

(a) the need in ozone of the water determines the relation flow to partial pressure of ozone in the gas;

(b) the dissolved oxygen necessary or useful in the treatment of the water determine the ratio of the partial flow to the pressure at the interface;

(c) the whole is established with taking into account the respective costs of the transfer of oxygen through the nitrogen bleed-off and generation of ozone at the desired concentration.

The recycling of the gas in the low-pressure technique does not necessitate extra precautions related to the resistance of the materials compared to those used currently in ozonization: stainless steel, copper, bronze, aluminium and cast iron. One must avoid the use of alloys rich in magnesium. So, the specifications will mainly take into account the transport of oxygen: smooth welding and exclusion of organic products (oils, lubricants, sealings, etc.).

In the high-pressure technique a compression of the ozonated gas to be injected is necessary to allow the diffusion in water under pressure. This phase is particularly delicate for what concerns the tightness of the gas circuits. In practice only the compressors with liquid water sealing in stainless steel of a high quality (AISI 316, 318 or similar) are usable. Their operating cost is high and is to be charged to the system considered as a whole.

Mainly for the ozonization of raw water or of sewage water it must be controlled that the recycled gas remains under the lower limits of explodability in case of degasification of organic impurities. No practical problem of this kind has been met in the existing plants.

REFERENCES

1. W. J. Masschelein, G. Fransolet, J. Genot, and R. Goossens, *T.S.M.-L'eau*, **71,** 385 (1976)
2. M. Simon and H. Scheidtmann, *G.W.F.*, **109,** 877 (1968)
3. H. M. Rosen, *Water and Sewage Works*, **119,** 114 (1972)
4. E. Albrecht, *Dechema-Monogr.*, **75,** 343 (1972).
5. O. Leitzke and G. Grener, *Symposium I.O.A., Nice 1979*, p. 76
6. G. Uhlig, *Symposium Karlsruhe* (1978) p. 357 (Procédés d'oxydation appliqués au traitement de l'eau potable)

CHAPTER 5

Alternative methods for the synthesis of ozone

W. J. MASSCHELEIN

Compagnie Intercommunale Bruxelloise des Eaux

Just as in the case for natural ozone, the formation of this oxidant can be obtained by UV irradiation of air or of a gas containing oxygen.

At ordinary pressure wavelengths normally active are inferior to 200 nm (region of Schumann–Runge at ordinary pressure). At high pressure or in condensed phase, the wavelengths superior to 200 nm and up to 280 nm are also active (region of Herzberg at high pressure) (see Figure 1).

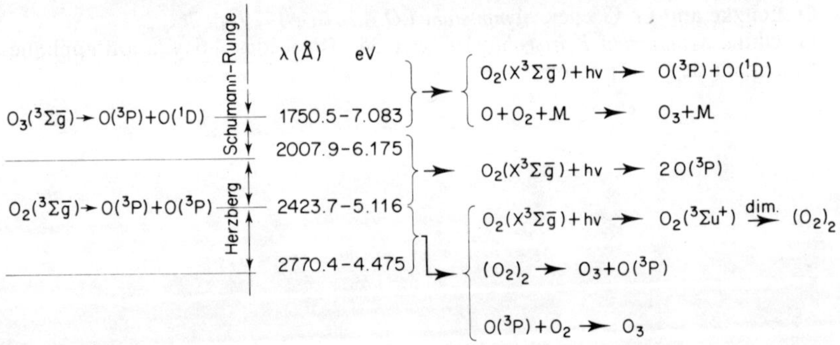

Figure 1. Photochemical formation of ozone (Herzberg range at high pressure; Schumann–Runge range at ordinary pressure

The way for the formation comprises the photochemical dissociation of oxygen into radicals ^1D and ^3P likely to initiate just as in the electronic discharge, the formation of ozone. In condensed phase, an activated dimer $(O_2)_2^0$ can also be formed.

The photochemical process has the advantages of its simplicity of putting into action. It only requires the use of a dry gas. The technique probably will not be applied to important plants for large-scale treatment of water because the ozone concentrations that can be reached at present are below 1 g O_3/m^3 of process gas and generally range 0.3 g/m^3 this for a specific energy consumption of 3 kWh/g. In extreme conditions, the energy yield attains 1.8 g O_3/kWh when operating in a gas phase and 3 g O_3/kWh when using liquid oxygen.

Electrical reduction of concentrated sulphuric acid is capable of forming ozone at a thermodynamic yield of 4 to 7 g/kWh. The thermal losses are high and the yield can be doubled by cooling of the reaction medium at −75°C. At present development, this process is not yet likely to be introduced in the schemes for water treatment although research on the subject is mentioned in the USSR.

Ozone synthesis by cheminuclear methods is a new technique which is not to be neglected in the future, seeing that potential use of waste products from nuclear power plants as Co^{60} is possible. In prototypes, the method enables to obtain ozone with a yield of 1.7 to 2.5 kWh/kg. The technique could be competitive with the classical methods from a production capacity of 200 kg O_3/day on. Particular aspects associated to this development are the following: necessity to operate with pure oxygen or an oxygen-enriched process gas at more than 90 % vol. O_2 so as to prevent excessive formation of nitrogen oxides. As a consequence the recycling of the off-gas is necessary. As the reactor is operated when surpressed (68 bar) an advanced technology is engaged in the process. A fractionation of the different gaseous irradiation products, e.g. noble gases, is imperative.

FURTHER READING

E. Briner, *AC Series*, **21**, 1 (1959)
M. Steinberg and M. Beller, *Chem. Eng. Progr. Symp. Ser.*, **66**, 205 (1970)
W. J. Masschelein, *T.S.M.-L'eau*, **77**, 177 (1977)

CHAPTER 6

Specifications and performance control for ozone generators

R. GERVAL

Syndicat des Communes de la Banlieue de Paris pour les Eaux

The specifications for ozone generators must be defined in a very practical way. Usually the first criterion for an ozonator is the nominal output, but this figure alone does not reveal much about the apparatus: the economic and operating conditions in which output was obtained must be specified. Many factors influence a generator's production, and all operating parameters should be taken into account. As to buying an ozonator: what controls are at the buyer's disposal for reception? What guarantees can he expect from the manufacturer to be assured of obtaining lasting promised performance with limited maintenance?

Figure 1 shows an ozone generator and its principal external connections: air circuit, or gas stream, cooling water circuit, electrical supply circuit. (Step-up transformer and inductance regulator or static voltage graduator.)

The first specification that must be imposed, to attain overall liability from the supplier, is to consider only offers for complete units which include delivery, installation and start-up of all air conditioning (pressurizing, cooling, drying), electrical supply, and ozone production equipment. Any offer of a partial nature should be shunned or examined critically. Anyway, in what concerns production and air treatment equipment, the manufacturer has complete freedom of design: choice of operating pressure, set-up of the cooling and drying stages. The supplier's sole obligation is that results are attained: these are the parameter specifications for entry into the ozonator outlined below.

PRESENCE OF HUMIDITY

The presence of moisture has a fundamental effect on the ozonator's output, and will determine its eventual reliability, as in the presence of moisture nitrogen oxides lead to nitric acid formation, which damages electrode surfaces and

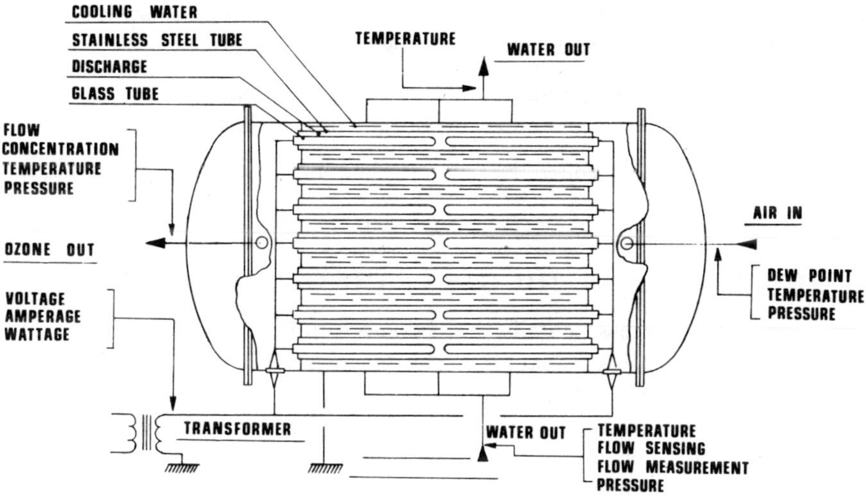

Figure 1. Ozone generator (horizontal tube technology)

dielectric tubes. The highest value to be guaranteed for the dew point should be −50°C.

Continuous measuring hygrometers should be installed between the dryers and the ozonators. For testing upon reception, measurements can be taken following the manual method (record of the temperature of condensation of air on cooling by carbonic snow in acetone).

Design must allow for constant dry-air flow, so that when an ozonator is stopped, it can be scavenged continuously, to avoid moisture entering the apparatus. Before voltage is applied, scavenging for half an hour should take place as well.

Moist-air backflow from the contact chambers must be avoided by a leak-proof non-return valve.

PRESENCE OF DUST

Dust may enter with the air from outside, or may derive from the adsorbent in the dryers. So as to avoid impairment of the generator's efficiency and to reduce maintenance, a filter should be located at the outlet of the dryer. Its efficiency should be 99% for particles of 1 micron, 98% for particles of 0.4 micron.

PRESSURE AT THE OZONATOR'S INLET

Pressure at the ozonator's inlet, usually determined by the profile of pressure losses in the air circuit and particularly in the contacting system, should be as close as possible to the following values, optimum compromise for ozone production in the horizontal tube technology:

(1) Ozone generator operating at 50 Hz: 0.55 relative bar pressure
(2) Ozone generator operating at 600 Hz: 0.8 relative bar pressure

This value will be controlled by a steel-plate manometer or by a micro-displacement capacitive cell.

TEMPERATURE OF THE GAS AT THE OZONATOR'S INLET

Any temperature increase entails the decomposition of produced ozone into oxygen. Yet in view of the substantial calorific power dissipated in the electrical discharge, the influence of the temperature of inlet air is secondary in the thermal make-up, at least within a reasonable range. The highest tolerable temperature is, temporarily, 35°C, after a regeneration cycle of the dryer. So as not to penalize the ozonator's cooling system, in regular service this value should not be above the temperature of cooling water used in the summer season, in the local conditions. The tender should specify the limits accepted in the proposal.

Temperature can be measured by means of a mercury thermometer or by a platinum-wire resistance probe.

AIR FLOW

Selection of a value for air flow derives in fact from the adopted work concentration in the gas stream. This point will be expanded upon below.

The flow-rate will be measured by a stainless steel orifice plate and a pressure-differential sensor. In the control upon reception, temperature and pressure should be measured at the same sampling point, so that the necessary corrections for measuring normal flow-measure conditions can be made.

TECHNICAL SPECIFICATIONS

Pipes upstream of the dryer should be of galvanized steel; those downstream of a material resisting ozone. They should be guaranteed to resist corrosion for at least 3 years. So as to avoid vibration and noise, they will be reckoned for a maximal speed of 10 m/s. The equipment should be pressure tested at a pressure 1 bar above normal operating pressure.

COOLING WATER CIRCUIT

The ozonator can be cooled with water deriving from the contact column and returning there after passage through the cooling circuit. This holds for generators operating at 50 Hz, if summer temperatures remain lower than 25°C. Otherwise the manufacturer must provide an accessory cooling system. The cooling circuit should be tested at an actual pressure of 3 bar. Thermometric sensors should be installed at the inlet and at the outlet of the generators. The maximum guaranteed temperature spread should be less than 5°C (the water-flow rate necessary to obtain these results is 100 l/h per dielectric tube in thin glass producing 18 g/h of ozone at nominal rate).

OZONE GENERATOR

The ozone generator will of course consist of materials resisting ozone. Before any operational use, the ozonator should be tested for leakage, as a closed system, by being put under pressure, at the chosen operating pressure for 12 h.

When the ozonator is put in operation, controls will aim to determine two values: ozone production expressed in grammes per hour, and the output of the generator expressed in watt-hours consumed per gramme of ozone produced.

Ozone production will be checked by means of air-flow feed rates and by the ozone concentration in the gas stream at the generator's outlet.

To record the output, power should be measured at the installation's low-tension display panel, so that only use by the ozonator's electrical equipment will be taken into consideration.

During these tests, other parameter values should be noted as well: air and water temperatures, pressure, dew point. To take the seasonal variations of cooling water temperature into account, these values should be determined periodically during the first year of guarantee. For every water temperature, results should be recorded on a graph illustrating the ozonator's production in relation to its power consumption, in the form of a network of curves corresponding to the various concentrations. The obtained results should be compared to analogous data furnished by the supplier. In view of uncertainties affecting the various measurements, tolerance of $\pm 5\%$ should be accorded.

All of these tests assuring satisfactory reception of the equipment should be undertaken by both the buyer and the supplier on a comparative basis.

CHAPTER 7

Principles in supplying electrical energy to ozonators

R. Mostin

Compagnie Intercommunale Bruxelloise des Eaux

GENERAL

Ozone production relies upon the use of a high-voltage alternative current of 6 to 20 kV applied to the gas-cell system which constitutes a condensor. Up to a limit value which depends on the properties and physical characteristics of the process gas the ozone production is proportional to the input of electrical power in the ozonator. This power input is expressed by the relation of Manley:

$$P = 4f\,Va\,[(Vo - Va)\,Cd - VaCa]$$

wherein:

f = current frequency in Hz
Va = voltage applied to the air (or gas) gap
Vo = extreme value of the applied voltage
Cd = capacity of the dielectric
Ca = capacity of the air (or gas) gap

From the electrical point of view an ozonator is equal to a capacity in which a single-phase high-voltage current is discharged. The production varies in function of the frequency and the voltage applied to the system. At present, in most existing systems, the production is varied and set at the operational value by changing the voltage applied; however it is also feasible to action on the frequency while maintaining a constant voltage.

EQUIPMENTS OPERATED AT INDUSTRIAL FREQUENCIES

To connect the ozonators to the electrical power supply system, whatever its voltage is, it is necessary to level down its value to a conventional low tension e.g. 220 V, 380 V so as to operate with a conventional regulating equipment (see

Figure 1). However, when the power input of the ozonators is important in comparison to the capability of the power supply, the output must be equilibrated among the three phases of the electrical current. To obtain this, the total capacity is divided between three ozonators which are operated simultaneously. The lower limit is to be set at the voltage enabling the minimal production of three ozonators (e.g. 11 kV for the classical horizontal tube ozonators).

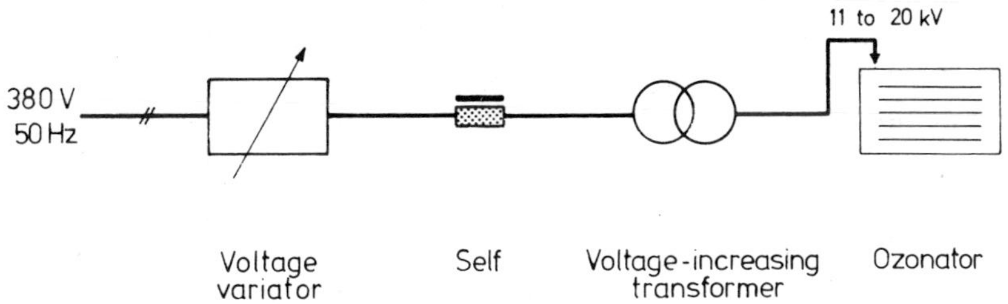

Figure 1. Conventional regulating system

To reduce the production capacity still further and maintain an equilibrated output from the different phases, a Scott-autotransformer is often set in. This equipment enables transformation of triphasic voltage into diphasic voltage and consequently operates only two ozonators at an equilibrated output rate.

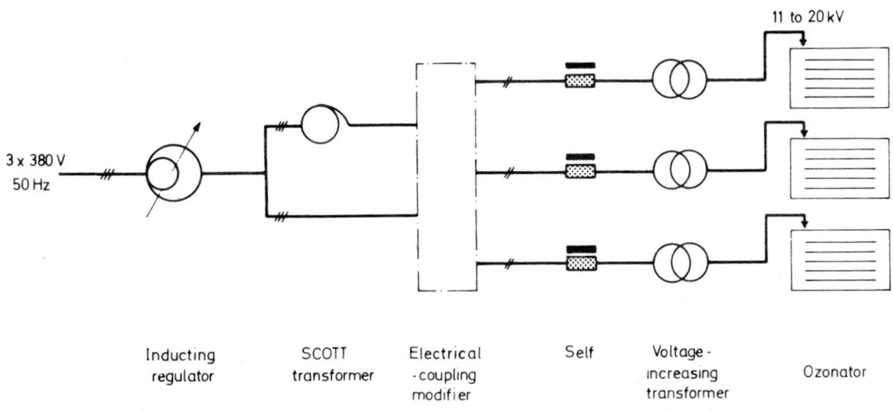

Figure 2. Principle of electrical coupling of the ozonators

To set the voltage at the necessary value when starting from the nominal voltage of 220 V or 380 V several classical means can be used.

(a) Inductive regulator (self), when the current is triphasic. This equipment is similar to an auto-transformer with a mobile primary coil *vs.* the secondary.

The voltage is varied through the respective positioning of the two coils.

(b) Auto-transformer with a slide contact. This device, which is simple and not very expensive, enables one to modify the voltage at each current phase by varying the number of wires in the secondary coil of the auto-transformer. Drawbacks are those associated with mechanical contacting systems which degrade through friction and spark effects.

(c) Voltage gradation system. In this technique thyristors are placed in the monophasic current. They are set up head-to-tail, and action is on the starting time. The system delivers an alternative tension at an effective value that can be regulated between zero and the full value. This simple and solid gradation system is of a full-electronic construction.

In all systems a monophasic transformer increases the low voltage up to the necessary value enabling the effluvation, i.e. 11 to 20 kV. The construction of the high-voltage transformer must be well adapted to be set in into the ozonator circuits in which important harmonic currents are developed. The power factor (cos φ) of ozonators which are capacitive components, varies between 0.3 and 0.5 according to changes in setting. To limit the intensities a compensating-self system can be set in so as to obtain a value of cos φ of about 1 at the maximal charge of the ozonator.

EQUIPMENTS OPERATED AT MEDIUM AND HIGH CURRENT FREQUENCIES

Here also the basic source of energy is the low-voltage (220 or 380 V, 50 Hz) current. Medium frequencies, e.g. 200 to 2,000 Hz, can be obtained either through a rotating set or through a semiconductor-based frequency converter. The latter technique which is the most economical is actually still under consideration. The equipment is composed of a current-redressor and a subsequent ondulator producing the desired frequency. Intermediately the circuit is equipped with a self to level out fluctuations in the redressed current.

The ondulator is designed to feed an ozonator: an oscillating circuit with an inductance coupled to the internal capacity of the ozonator. To equilibrate the

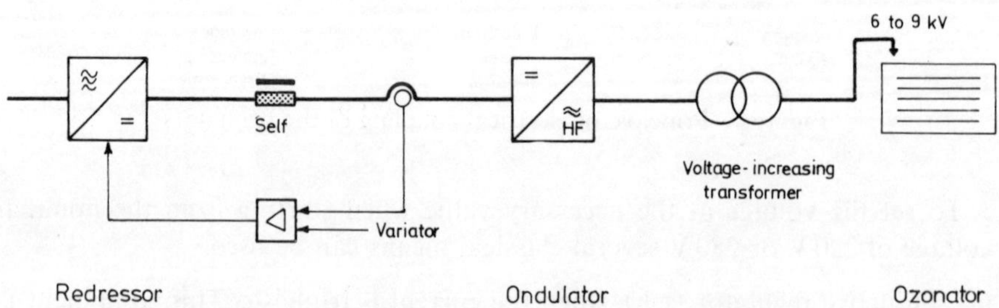

Figure 3. Thyristor-frequency variator coupling

power taken from the triphase current source, a redressor is incorporated to obtain the intermediate direct current. As the ozonators are fed at given values of the frequency, e.g. 200, 600 or 1,000 Hz, etc., the setting is operated by change in voltage applied to the ozonator.

The adjusting of the voltage is obtained by acting on the start–stop command of the thyristors incorporated in the redressor. Whence a DC voltage of changing value feeds the ondulator which converts the input into an alternative current of variable voltage and of a fixed frequency as determined by the characteristics of the ondulator. In this application the monophasic (voltage) transformer must be adequate for the chosen frequency. Its function is to increase the voltage to the value necessary to obtain the discharge. Usually the value adopted in this application is lower (e.g. 6 to 9 kV), than that in the equipment based on the direct use of the industrial frequency. Thus the risk of piercing the dielectrical material is lowered and, here also, it must be considered that the ozonator is submitted to important harmonic currents. By constituting the oscillating circuit a compensating effect of the reacting energy is obtained in the whole power range applied and an over-all value of $\cos \varphi$ of about 1 is obtained.

Alternative arrangements, in which the medium or high frequency is varied to obtain the production needs, are in present development.

Some of the constructors consider the frequency conversion techniques suitable for ozonators of high production capacity, i.e. when the size of the classical 50 Hz technology is to be increased strongly. Others have constructed and marketed medium-frequency units for small production capacities. Arguments advanced are: reduction in size, regular and equilibrated loading of the phases of the distribution net, less high operational voltage and correspondingly less risks of piercing the dielectrical material and the facility of electronical regulation of the systems.

It appears, however, that in the present state of development, at equal production (capacity), the electrical energy consumption of ozonators operated at higher–(e.g. 600 Hz)—frequency is 10–15% higher than the corresponding needs with 50 Hz current.

ELECTRICAL SECURITIES IN OZONIZATION PRAXIS

Classical electrical safety equipment is insufficient for ozone generation systems. The minimum requirements are to incorporate a relay system with instantaneous disconnection capability at maximum allowed current intensity of the system and set at a value near the maximum allowed nominal current intensity, e.g. safety level at 80–85% of nominal maximum current.

A typical accident in an ozonator is the piercing and breakage of a tube. This has, as a consequence, a local short-circuiting and all the tubes mounted in parallel are submitted to a sudden increase in current, from which more general damage may result.

The general relay-protection for maximal current-intensity does not allow a selective protection to each element: when one tube fails the whole production capacity breaks down. Therefore large ozonators are equipped with specific relay or fuse systems per part of an ozonator, e.g. 1/4, 1/3, 1/2. Other constructors propose an individual fuse system per tube so as to enable the continuation of the production when one or several tubes are out of work and to undertake the maintenance repairs later on. To be efficient the fuse systems must be set at well-calibrated values so that no weak links may cause cascade-type sequential fuse-ruptures which in turn may eventually momentarily increase the global current intensity in the remaining tubes beyond the critical level. Therefore the alternative still remains not to invest the amounts necessary for an individual-tube protection considering that in normal operation the breakages occur within the limits of the normal maintenance inspection and repairs.

FURTHER READING

Y. Faes, Ozoneurs: leur théorie et application dans technique nouvelles à semi-conducteurs à leur alimentation. *Rev. Gén. Electr.*, **84,** 1 (1975)

CHAPTER 8

Generation systems

Trailigaz ozone generator technology

P. CHAPSAL

Trailigaz

In its elementary form 'an electric discharge' type ozone generator has a minimum of two conducting electrodes maintained opposite one another in a manner such as to leave a uniform space between them. Into this space is introduced a thin solid dielectric, in general of glass, with a high permittivity and an area identical with that of the electrodes.

Ozone is produced when the space between the dielectric and the electrodes through which a dry gas containing oxygen passes, becomes conductive as soon as a sufficiently high AC current is applied. The assembly, from the electric point of view, is a capacitor with two dielectrics associated in series—one is a gas and the other a solid. An industrial generator developed recently by Trailigaz comprises concentric cylindrical electrodes provided with a centring arrangement which enables a very uniform discharge space to be obtained.

The metal parts are made of 316 L grade stainless steel and the dielectric elements of special, low-loss 'electronic' quality glass. The electrodes forming a 'heat-exchanger' type bank of tubes, as well as all the metal parts constitute a general mass connected to earth, which eliminates all danger of electrocution. The active surfaces of the electrodes are smooth and manufacturing tolerances of the tubular elements of the bank of tubes, as well as the dielectrics, are extremely rigorous.

The glass dielectric tubes are hot metal-coated internally and are annealed after shaking to avoid the occurrence of any mechanical stress in the material.

An AC current is applied to the metal coating of these dielectrics. The connections for transferring electricity to the metal coats on the dielectrics are connected to one another and to a strictly airtight and insulated collector system, so as to avoid gas leaks as well as accidental contacts. The appropriate cylindrical form of the body and the 'autoclave' type closure cover provide complete sealing.

A generator can comprise several hundred electrodes without giving rise to any constructional difficulty or take up very large space. The range of Trailigaz generators enables ozone to be produced at various output rates and so can satisfy requirements ranging up to several hundred kg O_3/h. A 'low' range is in fact standardized and includes, assembled on a single frame, all the essential parts of a complete ozone production unit for the treatment of water and other applications. The complete unit, checked and tested in the factory, therefore requires only simple electrical connections, cooling water and air or other gases containing oxygen. The generators also include suitable apparatus for drying air or other gases down to a dew point of between −50°C and −80°C.

The preliminary drying of gases is necessary to obtain an acceptable yield of ozone and avoid the formation of considerable amounts of nitrogen in the ordinary case when air is used. These compounds are deposited, particularly on the active parts of the electrodes, which require frequent cleaning if the drying of the air is not sufficiently thorough.

The drying equipment makes uses of processes involving adsorption on activated alumina. Regeneration is entirely automatic and is obtained by flushing through hot air.

Electric power is supplied through a single-phase special transformer with a Scott type enabling the phases on the network side to be balanced. The sinusoidal AC current is applied to these transformers with a suitable amplitude by means of conventional variators (induction regulators, or thyristor systems).

Generally the normal low frequency is used, namely 50 or 60 Hz. In some circumstances, taking their price and yield into account, static frequency converters, of which the technology is now conventional, can also be adapted. These devices enable the production of ozone of a given unit to be increased under identical physical–chemical and thermodynamic conditions.

The dielectric capacity of the ozonator being variable, the system is a great generator of uneven-number harmonics (80 % harmonics of the order 3). The current absorbed, which is not sinusoidal, can cause disturbances: a power factor unacceptable to the distribution network, which adversely affects the durability of the dielectrics. Such disturbances increase as the size of the ozonator increases. But the Trailigaz ozonator has been designed to avoid these defects and provide absolutely reliable operation. These electrical arrangements, together with the manufacture of generator elements (tubes) to particularly fine tolerances, enabling the expected performances to be obtained: guarantees covering ozone production and specific energy consumption, which are perfectly reproducible from one apparatus to another, whatever its capacity.

Performances vary according to the quality on the composition of the gas that is used, e.g. its water and oxygen content and to the following physical–chemical conditions: Pressure and temperature of gas, content of ozone in gas, temperature and quality of cooling fluids.

As an indication, for 50 Hz, and for the usual concentrations of ozone of 12 to

$20\,g\,O_3/Nm^3$ of air, with cooling water at 15°C, the specific consumption of electrical energy varies between 13 and 18 KWh/kg O_3 produced, following also the possible adjustment range of the ozonator. Under these conditions, the corresponding production of a large-capacity Trailigaz ozonator attains 13 to 15 kg O_3/h. This performance is appreciably increased with 600 Hz power supply, production attaining 35 kg O_3/h, if the current is supplied through a frequency converter. The technology also supports the use of oxygen-enriched process gas and of gases containing carbon dioxide.

Figure 1. First section of six ozonators: City of Montreal (Canada)

Degrémont's ozonators

R. LOUBOUTIN

Degrémont

Degrémont's ozonators are of the horizontal tube type. They consist of a cylindrical horizontal tank made of stainless steel closed off at each end by a dished bottom fitted with one or more glass sights. The tank contains a tube bundle welded to end places, the peripheries of which are welded to the tank shell. These tubes and plates are also made of stainless steel.

The cylindrical shell of the ozonator, the end plates as well as the outer walls of the metal tubes delimit a central compartment in which flowing water cools the tube bundle.

Each metal tube surrounds a dielectric in the form of a cylindrical glass tube closed off at one end. The inner side of the tube wall is covered with an electrically conducting metal coat. This metal coating provides a high-voltage electrode wired to a single-phase AC source, whereas the stainless steel tube acts as a second electrode grounded via the casing of the ozonator.

The gas used (dry air or oxygen) flows in the annular space marked off by the metal tube and the dielectric tube.

The glass dielectric provides an equipotential surface at the boundary of the discharge gap. It must have good dielectric strength as well as high permittivity in order to minimize its resistance to the flow of alternating current.

Since dimensional uniformity of the annular space in which the electric current is created is essential for correct operation of an ozonator, the diameter and thickness of the dielectrics equipping the Degrémont units are calibrated. In addition, the width of the flow gap determined by the distance between the inner wall of the metal tube and the outer wall of the dielectric tube is kept constant by patented stainless steel centring devices.

Electric power is supplied to the high-voltage electrodes by special patented connectors. These connectors, which closely follow the shape of the equipotential surfaces and prolong the metal electrode to a certain extent, avoid the appearance (above a certain voltage) of the spark phenomenon due to the peak effects at the end of the conducting metal layer.

These connectors may also be provided with a high-voltage fuse which, in case of an exceptional accident involving the dielectric tube and resulting in a local short circuit, eliminates this dielectric tube from the electric circuit without stopping the generator and making it necessary to immediately replace the defective dielectric.

Degrémont's industrial ozonators are produced in numerous standard sizes. Figure 2(a) shows the type CL ozonator in which each metal tube of the bundle contains a single dielectric tube, whereas the type GL ozonator shown in Figure 2(b) has two dielectric tubes mounted 'head-to-tail' in each metal tube. Degrémont thus produces units containing up to 600 dielectric tubes in a single shell, the unit production rates of which can reach several score kilogrammes of ozone an hour.

The maximum ozone production rate for a given ozonator depends on a certain number of parameters and, in particular, the type of gas used (air, oxygen-enriched air or pure oxygen), the concentration of ozone produced in the vector gas and the frequency of the power supply current.

If electricity is supplied at industrial frequency, the production rate of the

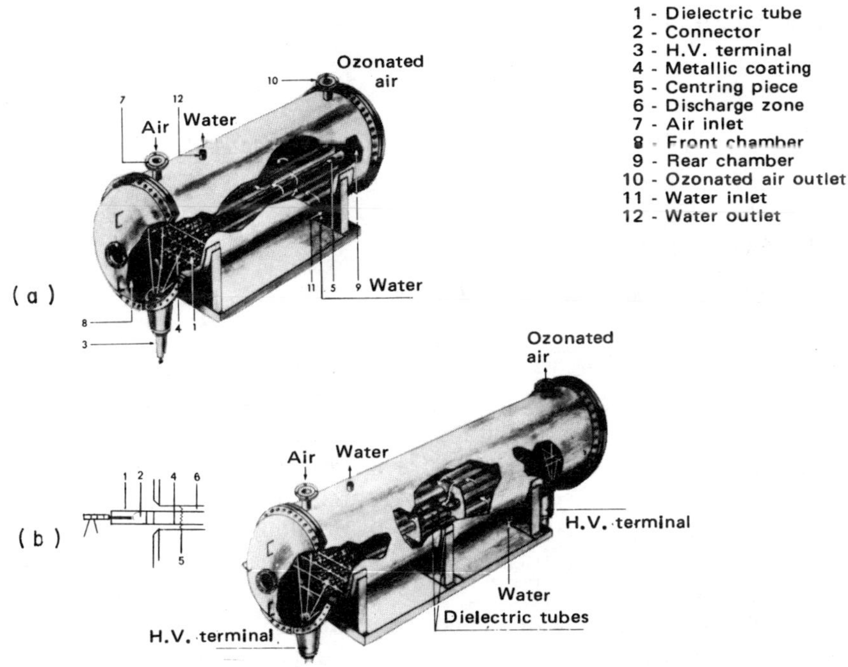

Figure 2. (a) Degrémont CL-type ozonator; (b) Degrémont GL-type ozonator

ozonizer is adjusted by modifying the value of the voltage applied to it. If the power is supplied at medium frequency (500 to 2,000 Hz range), making it possible to practically double the maximum output for a given unit, the production rate is adjusted by changing the thyristor conducting time of the thyrotron inverter generating this medium frequency.

An incremental controller with microprocessor, specially developed by Degrémont for control of its ozonator, makes it possible to automatically control the ozone production rate by the flow rate of the water to be treated and the value of residual ozone in the treated water.

The power consumed by the Degrémont ozonators depends on the above-mentioned parameters. Two typical values for a concentration of 1 % in air and 2 % in oxygen are 16 kWh and 8.3 kWh respectively per kg of ozone produced at the nominal load of the ozonizer. Naturally, these values decrease when the load decreases.

Finally, it should be mentioned that, in addition to the type CL and GL ozonators, Degrémont produces laboratory model ozonators providing up to 20 g/h of ozone with dry air as well as one-piece cabinet-mounted ozonators consisting of unit ozone generators with a single dielectric. When placed in parallel, these ozonators can produce 200 g/h of ozone from air at a frequency of 50 Hz.

The Litha ozonator

W. J. MASSCHELEIN

Compagnie Intercommunale Bruxelloise des Eaux

In much reduced scale equipment, pilot plants and laboratory work an important problem can be to dispose of a small apparatus, easy to operate for the synthetis of ozone. The Litha ozonator still constructed by piece is largely adjustable according to the demand and corresponds to this need. The ozonator is operated without cooling except that by the process gas. It is fed directly from any conventional source of electrical energy of single-phase 220 V, 50 Hz, other options being possible.

Originally developed to treat and purify–disinfect industrial gases and atmospheres, the basic element of the ozonator is its dielectrical plates in mica calibrated at 0.2 mm thickness. The electrical discharge is obtained at less than 2.9 kV (regulated by a transformer incorporated into the system) and at an 'induced frequency' lower than 400 Hz between the mass-electrode and an active electrode of about 30 cm^2 in stainless steel sandwiched within the dielectrics. Other essential characteristics are:

average electrical field: 25,800 V/cm;
apparent current density: 0.23 mA/cm^2;

nominal residence time of the gas in the discharge zone: 0.23 s. The ozonator can be operated in the pressure zone of 0 to 1 bar.

The nominal production of one unit chamber with double gas space attains up to 100 mg O_3/h when using dry air as effluvated gas. The general characteristics of the ozonator enable one to obtain a production of 2.5 times more ozone when using pure oxygen instead of air. The thermodynamic yield is in the region of the performances obtained with the best ozonators, i.e. a production of 3.3 and (more particularly) 8.5 molecules of ozone per 100 eV respectively when using air or oxygen.

In nominal operating conditions the ozone concentration is of 4 g/Nm^3 and of 10 g/Nm^3 respectively when using air and oxygen. When giving-up part of the yield and admitting some heating of the process gas under the maximal allowed limit of 60°C it is possible to establish and to hold on increased stationary concentrations in ozone by diminishing the gas-flow. The maximal concentration can thus attain 35 g/Nm^3 and 85 g/Nm^3 when starting from air and oxygen respectively.

The total production capacity of any given ozonator can be adjusted by parallel arrangement of several double-spaced discharge zones and by placing two or several of these configurations in series. In this sense the most used model for laboratory work is the 256-type with twelve double gas chambers (Figure 3) involving an arrangement of two groups in series each composed of six discharge gaps set up in parallel. The nominal production capacity with air is of 2.2 g O_3/h. Equipments are constructed and assembled to cover the range of production capacities per unit installed from 0.1 to 2.2 g O_3/h when using air.

According to our experience, this equipment is of great suppleness of use and being of another order of the ozonators for large treatment plants it answers perfectly to the needs of research work and development of the principles and the practical execution of ozonation.

Figure 3. Litha ozonators with twelve gas-spaces

Furthermore a double open chamber set with a nominal production capacity of 5 g O_3/h fed with a double electrical circuit and built-in cooling and drying is now produced for higher demands. The open-chamber technology favours the cooling. The operation is controlled with a pressiostat at less than 1 bar and the double electrical 220 V power feed secures the operation. Up to six sets in parallel have now been operated simultaneously thus enabling a 30 g/h capacity with air. The elementary electric gaps are of the same principle as for the laboratory equipment.

Mather and Platt Capenhurst ozonator

R. K. W. PALMER

Mather & Platt

The principle of ozone generation by applying high alternating voltage discharge to dry air, passing between metal and dielectric electrodes, is well known. To increase ozone generation for a given electrode configuration, power density must be increased by raising the applied voltage and/or frequency.

On conventional ozonators, only the earth electrode is directly water-cooled. As power density is increased, overheating of the high-tension electrode (not directly cooled) causes thermal decomposition of the ozone in the air gap, thus reducing efficiency.

The 'Capenhurst' system (which is subject to world-wide patents) directly water-cools both high-tension and earth electrodes allowing high power densities (frequencies up to 3 kHz), and an 85% reduction in electrode volume compared with conventional ozonisers (Figure 5). The corresponding reduction in floor area can substantially reduce ozonization plant building costs.

Direct cooling of both electrodes maximizes the amount of heat removable from the discharge, minimizing power for ozone production.

The 'Capenhurst' high-frequency system (Figure 6) directly switches high voltage, giving greater control of power, without the losses inherent in low voltage variable frequency systems.

Since control of ozone production is by frequency, ozonator operating voltage can be maintained at an optimum value to minimize dielectric stress and losses.

Stepless variation of electrode power over a wide range is achieved, simply and efficiently, by electronic control of inverter frequency which gives great operational flexibility. Since both electrodes are directly water-cooled, very low air flow-rates are acceptable, which permit ozone generation at high

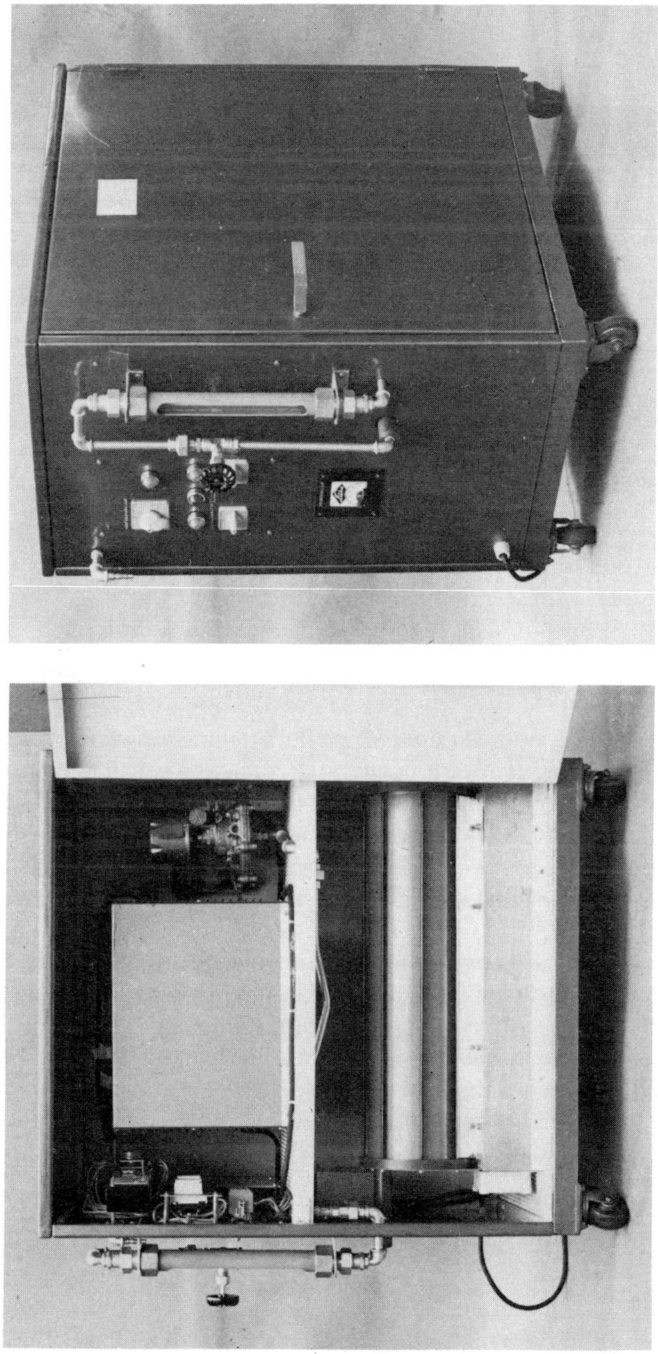

Figure 4. Litha open-chamber ozonators for general uses. (Producer: Litha, rue des Vennes 273, 4000 Liège, Belgium)

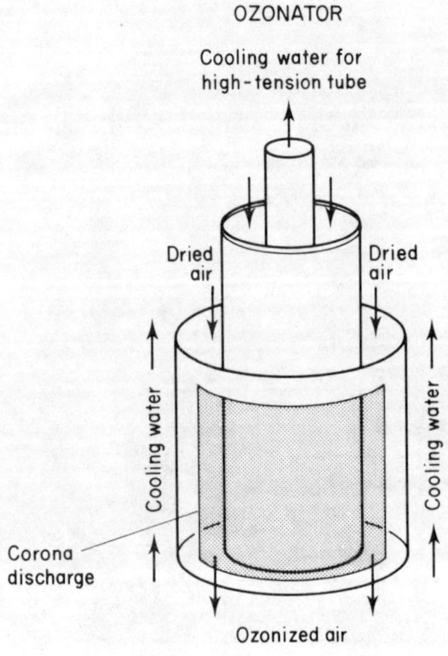

Figure 5. Cooling principle of the Capenhurst ozonator

concentration. Using both features, 'Capenhurst' units can give a much greater range of ozone production (turn-down ratio), even at a constant concentration.

The design of the cell also allows operation at any of the pressures normally required for ozone application systems (0 to 10 m w.g.).

The 'Capenhurst' ozonator system is based on the use of standard electrode modules (nominal capacity of 1 kg/h from air or 2.5 kg/h from oxygen) grouped together with a power module to form an ozone generator block of required output. This system has many maintenance advantages allowing operating costs to be minimized. A cell can be removed, and a replacement cell quickly fitted, allowing maintenance to be undertaken with minimum ozonator down-time. Individual cell modules are also easily serviced. Electrode and dielectric tubes can be exchanged in under an hour.

A similar modular concept is used for the electrical components within the inverter and power unit.

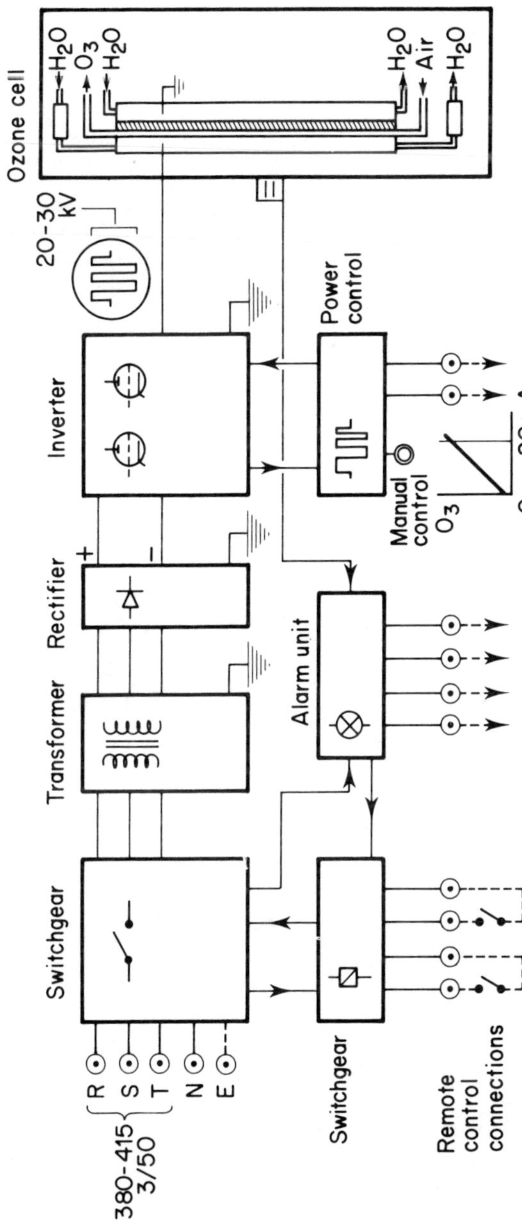

Figure 6. High-frequency control system

2. FUNDAMENTAL PRINCIPLES OF THE OZONIZATION OF WATER

CHAPTER 9

Theoretical aspects of ozone transfer into water

M. ROUSTAN* and J. MALLEVIALLE†

**INSA Chimie; †SLEE*

In order to examine the effectiveness of the ozone–water contactor system, it is essential to determine experimentally the transfer coefficient of ozone in water. These investigations should not be confused with those concerning the rate of breakdown of certain organic compounds. We shall begin by dealing with the case of the transfer ozone (physical absorption) where there is no chemical reaction of the ozone with the substances contained in the water. We shall then go on to consider the case where the ozone reacts in liquid phase with the substances, at different reaction rates.

PHYSICAL ABSORPTION

By means of the double-film theory it is possible to predict the rate of transfer of a substance from one phase to another phase (see Figure 1).

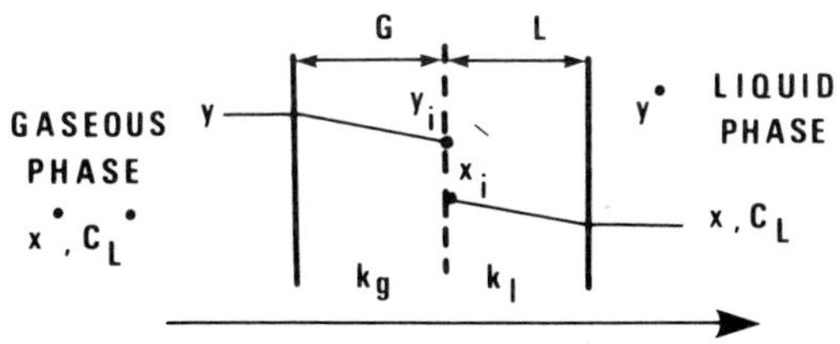

Figure 1

The equation of transfer may be written:

[amount transferred per unit time] =

[transfer coefficient][exchange area][exchange potential]

i.e., on the gas side:

$$N_A = K_G S(y - y^{\bullet}) = k_g S\,(y - y_i) \tag{1}$$

on the liquid side:

$$N_A = K_L S(C_L^{\bullet} - C_L) = k_l S(C_i - C_L) \tag{2}$$

where:

$C_L^{\bullet}$ is the concentration of the substance A in equilibrium in the liquid phase with a gas of composition $y^{\bullet}$

$y^{\bullet}$ is the molar fraction of the substance A in equilibrium in the gas phase with a liquid whose concentration is $C_L^{\bullet}$

C_i is the concentration of the substance A on the liquid film side of the interface;

y_i is the molar fraction of substance A on the gas film side of the interface;

k_l is the film transfer coefficient on the liquid side;

k_g is the film transfer coefficient on the gas side;

S is the exchange area;

N_A is the amount transferred per unit time;

K_L is the total transfer coefficient on the liquid side;

K_G is the total transfer coefficient on the gas side.

If at equilibrium there is a simple relationship between y and x (where the equilibrium curve can be taken as a straight line $y = mx$), the equations (1) and (2) may be combined to give:

$$\frac{1}{K_L} = \frac{1}{k_l} + \frac{1}{mk_g} \qquad \frac{1}{K_G} = \frac{1}{k_g} + \frac{m}{k_l}$$

The coefficient m is in fact Henry's constant He, of values indicated in Table 1 for ozone.

Since ozone has a low solubility, the entire resistance to transfer can be considered as residing in the liquid film, therefore

$$\frac{1}{K_L} \simeq \frac{1}{k_l} \quad \text{and } C_i = C_L^{\bullet}.$$

Table 1

	T^0	5	10	15	20	25	30
$y = He.x$	He	2180	2480	2880	3760	4570	5980

y = partial pressure (atm)
x = molar fraction in the liquid phase
He is given in atm/molar fraction

The equation (2) becomes:

$$\boxed{N_A = k_l S(C_L^{\bullet} - C_L)} \tag{3}$$

In a gas–liquid exchange system, it is often difficult to determine the exchange area S. It is then preferable to express S in terms of volume of liquid V, i.e. $a = S/V$, the specific exchange area (L^2/L^3).

Equation (3) becomes:

$$\boxed{N_A = k_l a(C_L^{\bullet} - C_L)V} \tag{4}$$

The coefficient $k_l a$ (T^{-1}) can be determined in a number of ways:

(a) In transient conditions this involves determining the rate of increase in the concentration of dissolved ozone in the previously ozone-free liquid phase.
(b) In steady conditions it requires the providing of a contactor with a steady supply of water not consuming ozone and a flow of ozonized air. By carrying out a mass balance in the gaseous and liquid phases it is possible to determine the value of the transfer coefficient if the flow models in the liquid and gaseous phases are known.

The coefficient of diffusivity Do_3 of the ozone in the water can be calculated using the correlation of Wilke and Chang:

$$\text{at } 20°\text{C}, \; Do_3 = 1.74 \times 10^{-5}\,\text{cm}^2/\text{s}.$$

If the ozone is dispersed in the form of bubbles, the ozone-water dispersion may be characterized by:

(a) the gas retention: h
(b) the bubble diameter: d_B
(c) the specific exchange area: a

related as:

$$a = \frac{6}{d_B}\frac{h}{1-h}$$

By determining d_B and h it is possible to find a and hence k_l from the coefficient $k_l a$.

A few correlations are available enabling the film coefficient k_l to be calculated *a priori* for bubbles in free ascension:

Higbie: $$k_l = 1.13\frac{\sqrt{D}}{t_C} \qquad Sh = 1.13\,Re^{0.50}\,S_c^{0.50}$$

Hugmark: $$Sh = 2 + a'\left[Re^{0.484}Sc^{0.34}\left(\frac{d_B g^{0.33}}{D^{0.66}}\right)^{0.072}\right]^b$$

t_C is the contact time between a liquid and a gas, assumed to be equal to the time

the bubble takes to travel a distance equal to its diameter d_B. $t_C = d_B/U_t$ where U_t is the limiting upward velocity of the bubble. The Higbie model is applicable if d_B is greater than 2.5 mm and if the bubbles rise freely and separately.
In Hugmark's formula:

$b = 1.61$ $a' = 0.061$ for separate bubbles
$a' = 0.0187$ for 'packets' of bubbles

$Re = \frac{U_S d_B}{v}$ U_S = velocity of the bubble (or packet) with respect to the liquid.

Table 2 gives values of k_l calculated using the Higbie and Hugmark formula assuming $U_S = U_t = 0.25$ m/s.

Table 2. (assumption $U_t = 0.25$ m/s)

	k_l (cm/s)		
d_B (mm)	Higbie	Hugmark $a' = 0.061$	Hugmark $a' = 0.0187$
2.5	0.0471	0.0527	0.0162
3	0.0430	0.0517	0.0159
4	0.0373	0.0501	0.0154
5	0.0333	0.0489	0.0150
6	0.0304	0.0480	0.0148

ABSORPTION WITH CHEMICAL REACTION

If the ozone reacts with certain organic constituents of the water (organic compounds, viruses), it is necessary to know whether this reaction increases the amount of ozone transferred. This is done by determining the acceleration factor E defined by:

$$E = \frac{k_{l'} \text{ with reaction}}{k_l \text{ without reaction}}$$

Assuming a first-order reaction rate, Danckwerts[1] gives:

$$E = \sqrt{1 + M}\,;\; M = \frac{D_{O_3} . k}{k_l^2}$$

Table 3 shows the effect of reaction rate on the acceleration factor E.

The factor E does in fact depend also on the hydrodynamics of the ozone–water contact system. Table 4 shows the order of magnitude of the reaction constant k for the factor E to be considered as equal to unity ($M \ll 1$).

A number of experiments have been carried out both on physical absorption and on absorption with a chemical reaction. The results obtained are summarized in Table 5.

Table 3

Very slow reaction	Very little ozone is consumed by the reaction $N_A = k_l a(C_L^* - C_L)V$	$E = 1$ if $M \ll 1$
Slow reaction	In certain conditions the reaction is sufficiently fast for $C_L \approx 0$ $N_A = k_l a C_L^* V$	$E \gg 1$
Fast reaction	All the ozone reacts in the liquid film – $C_L \approx 0$ $k_l a = \sqrt{Do_3 k} . a$ $N_A = C_L \sqrt{Do_3 k} . S$	$E > 1$ $M \gg 1$ $E = \sqrt{M}$

Table 4

k_l (cm/sec)	Type of contactor	k (s^{-1})	Order of magnitude of k to obtain $E \neq 1$
0.05	Bubble column	$\ll 140$	10
0.03	$2.5 < d_B < 5$ mm	$\ll 50$	5
0.02	Tank stirred	$\ll 20$	2
0.01	Bubble column	$\ll 5$	0.5
0.005	$d_B < 2$ mm	$\ll 1.4$	0.1
0.001	Packed column	$\ll 0.05$	0.005

Table 5

Physical absorption

Hill and Spencer	Bubble column $d_B = 2.5$ mm	$k_l = 0.02$ cm/s
Roustan, Mallevialle	Tank stirred mechanically	$k_l = 0.025$ cm/s
Augugliaro and Rizzuti	Descending film column	$k_l = 0.0067$ cm/s

Chemical absorption

Hill and Spencer — Bubble column, Virus disinfection

d_B (mm)	k_l (cm/s)	k_l (cm/s)	E (s)
6.2	0.047	0.014	3.3
3.3	0.032	0.014	1.7
2.6	0.030	0.021	1.4

Augugliaro and Rizzuti — Oxidation of an aqueous solution of phenol

pH	Phenol (m/l)	k_l' (cm/s)	k (s^{-1})	E (s)
4.66	80	0.0114	8.4	1.7
5.45	80	0.0301	51.8	4.5
5.61	80	0.0330	748	5
6	80	0.0470	184	7

REFERENCES AND FURTHER READING

V. Augugliaro and L. Rizzuti, *Chem. Eng. Sci.*, **33**, 1441–7 (1978)
P. V. Danckwerts, *Gas–Liquid Reactions*, McGraw-Hill (1970)
J. A. Heist, *Am. Inst. Chem. Engrs.*, pp. 456–7 (1974)
A. G. Hill and H. T. Spencer, *1st Symposium on Ozone for Water and Wastewater Treatment*. Washington D.C., December (1975)
J. Mallevialle, M. Roustan, H. Roques, Cebedeau, pp. 175–9 (1975)
Perry, *Chemical Engineer's Handbook*, McGraw-Hill (1980)
D. A. Richards and M. Fleischman, *2nd Int. Symp. on Ozone Technology*, Montreal, Canada, May (1975)
M. Roustan, Doctoral thesis. No. 809—Toulouse (1978)
C. Scaccia and H. M. Rosen, *Ozonews, Part 2*, **5**, October (1978)

CHAPTER 10

Stability of ozone in water

J. MALLEVIALLE

SLEE

The auto-decomposition of ozone in water has been studied by a number of workers with two different objectives: firstly, to establish the mechanisms of auto-decomposition as such; and secondly, to study the reaction kinetics of decomposition.

Summarizing all this research turns out to be very difficult, because the results are very disparate. In research of this type a number of phenomena are involved, and it is fairly difficult to evaluate the contribution of each of them:

(a) transfer of ozone from the gas phase to the liquid phase;
(b) relation between the partial pressure of the gaseous ozone and its solubility in an aqueous solution;
(c) mass transfer of the ozone dissolved to trace impurities present in the water;
(d) kinetics of the auto-decomposition of ozone in solution;
(e) kinetics of the oxidation by ozone of the impurities in the water.

Besides all such phenomena, these studies rely also upon the analytical methods used for determining the ozone. Since no ozone standard is available, the reference most frequently used is still the release of iodine in a solution of potassium iodide.

The solubility of ozone in water does of course depend on the values of the equilibrium constants in Henry's law. A number of values given in the literature are shown in Figure 1. Generally speaking these data were obtained in 'pure' water. It is obvious that the solubility of ozone is substantially influenced by the presence of impurities in the water, for example, alkaline salts reduce the solubility while neutral salts increase it.

Several workers studying the mechanisms of auto-decomposition of ozone have suggested different intermediate stages in the process. The chemical species most often involved are:

$$O_3 + OH^- \rightarrow O_2^- + HO_2$$
$$O_3 + H_2O \rightarrow H_3O^- + OH^- = 2\,HO_2$$

$$O_3 + OH^- \rightarrow O^-.O.O.OH$$
$$O_3 + 2OH^- \rightarrow O_2 + 2O^- + H_2O$$
$$O_3 + OH^- \rightarrow O_3^- + OH$$

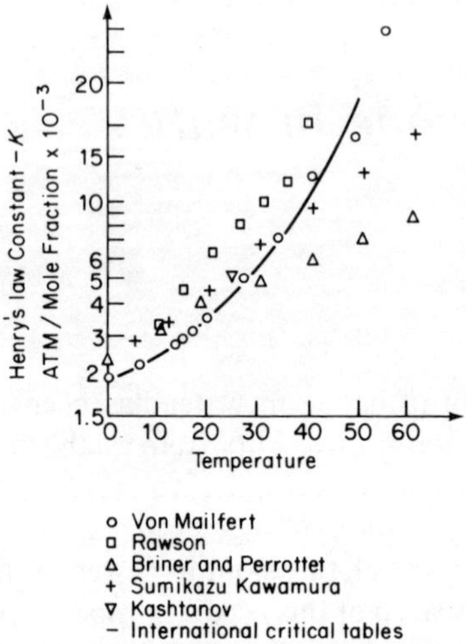

Figure 1. Solubility of ozone in water

Various reaction kinetics corresponding to different ranges of temperature and pH are given in the literature. A few examples are given in Table 1, which illustrates the disparity of the results.

The influence of certain parameters is generally recognized. For example, it is accepted that hydroxyl ions act as catalysts in the decomposition of ozone in an aqueous solution. However, considerable uncertainty persists in most of the results given in the literature and some fundamental research is still to be undertaken on the subject.

FURTHER READING

J. S. Roth, D. E. Sullivan and M. D'Ambroisi, Symposium on Advanced Ozone Technology, held 16–18 November 1977 in Toronto (Canada), 10A, USA
E. Moufang, Abstr. *J. Soc. Chem. Ind.*, **30,** 1210 (1911)
J. Weiss, *Trans. Farad. Soc.*, **319,** 668 (1935)
M. G. Alder, Ph.D., Dissertation, University of Utah (1949)
E. Abel, *Monatsh. Chem.*, **86,** 44 (1955)
V. Landi and L. Height, *J. Phys. Chem.*, **73,** 2361 (1969)
D. W. Gorbenko-Germanov *et al.*, *Chem. Abstr.*, **82,** 13921k, 139216m (1975)
M. Peleg, *Water Research*, **10,** 361 (1976)
J. Hoigne and H. Bader, *Chimica*, **29,** 20 (1975)

Table 1

Reference		pH range	Temperature range	Reaction order
Rothmund and Burgstaller	1913	2 → 4	0	2
Sennewald	1933	5.3 → 8	0	2
Weiss	1935	2 → 8	0	3/2
Adler and Hill	1950	1 → 2,8	0 → 27	1
Stumm	1954	7.6 → 10.4	1.2 → 19.8	1
Kilpatrick *et al.*	1956	0 → 6.8	25	3/2
Killpatrick *et al.*	1956	13 →	25	2
Rankas *et al.*	1962	5.4 → 8.5	5 → 25	3/2
Hewes and Davison	1971	2 → 4	30 → 60	2
Hewes and Davison	1971	6	10 → 50	3/2 → 2
Hewes and Davison	1971	8	10 → 20	1
Czapski	1968	10 → 13	25	1
Rogozhkin	1970	9.6 → 11.9	25	1
Merkulova *et al.*	1971	0.22 → 1.9	5 → 40	1 or 2

REFERENCES

1. M. G. Adler and G. R. Hill (1950) *J. Am. Chem. Soc.*, **72,** 1884
2. E. Briner and E. Perrottet (1937) *Helv. Chim. Acta*, **22,** 397
3. C. G. Hewes and R. R. Davison (1971) in: A critical review of kinetics of ozone decomposition in aqueous solutions, J. S. Roth *et al.* (1979) *Ozonews*, Part 2, Technical Paper Section, **7,** 7 (July)
4. International critical tables (1926). Vol. 3,257.
5. S. W. Kashtanov *et al.* (1975) *Zh. Prikl. Khim.*, **48,** 8.
6. S. Kawamura, (1932) *J. Chem. Soc. Japan*, **53,** 783
7. M. L. Kilpatrick *et al.* (1955) *J. Am. Chem. Soc.*, **78,** 1784
8. V. P. Merkulova *et al.* (1971), in: A critical review of kinetics of ozone decomposition in aqueous solutions, Roth, J. S., *et al.* (1979), *Ozonews*, Part 2, Technical Paper, Section, **7,** 7 (July)
9. M. M. Rankas *et al.* (1962), in: A critical review of kinetics of ozone decomposition in aqueous solutions, J. S. Roth *et al.* (1979) *Ozonews*, Part 2, Technical Paper Section, **7,** 7 (July)
10. A. E. Rawson (1955) *Waste Water Eng.*, **57,** 102
11. G. I. Rogozhkin (1979) in: A critical review of kinetics of ozone decomposition in aqueous solutions, J. S. Roth *et al.* (1979) *Ozonews*, Part 2, Technical Paper Section, **7,** 7 (July)
12. V. Rothmund and A. Burgstaller (1913) in: A critical review of kinetics of ozone decomposition in aqueous solutions, J. S. Roth *et al.* (1979) *Ozonews*, Part 2, Technical Paper Section, **7,** 7 (July)
13. K. Sennewald (1933) in: A critical review of kinetics of ozone decomposition in aqueous solutions, J. S. Roth *et al.* (1979) *Ozonews*, Part 2, Technical Paper Section, **7,** 7 (July)
14. W. Stumm (1954) in: A critical review of kinetics of ozone decomposition in aqueous solutions, J. S. Roth *et al.* (1979) *Ozonews*, Part 2, Technical Paper Section, **7,** 7 (July)

15. H. Von Mailfert (1894) *Compt. Rend.*, **119**, 951
16. J. Weiss (1935) *J. Trans. Farad. Soc.*, **319**, 668

CHAPTER 11

Reaction pathways and kinetics of reactions of ozone with compounds dissolved in water

J. HOIGNÉ

Federal Institute for Water Resources and Water Pollution Control

When ozone is used to oxidize a solute in water, one of three possible pathways determines the rate at which the chemical reaction proceeds. These different pathways are:

(a) direct reaction of the solute with ozone;
(b) oxidation of the solute by $OH^{\cdot}$ radicals formed from decomposed ozone;
(c) solute-induced oxidation reactions.

DIRECT REACTION OF OZONE WITH THE SOLUTE

Reactions of this type may be represented by the generalized scheme:

$$M + O_3 \rightarrow \rightarrow \cdots \rightarrow M'_{oxid.} + M''_{oxid.} + \cdots$$

where M is the solute and $M'_{oxid.}$ and $M''_{oxid.}$ are quasi-final oxidation products which are only slowly or not at all further oxidized by ozone.

The rate of reaction of M is first-order with respect to the concentrations of both O_3 and M. In a plug-flow or batch-type reactor and under pseudo-first order conditions the integrated rate law for the 'elimination' of M is:

$$-\ln([M]_t/[M]_0) = k_M[\overline{O_3}]t \qquad (1)$$

where $[M]_t$ is the concentration of M residual at time t: $[M]_0$ is the concentration of M at time zero; k_M is the second-order rate constant of M reacting with O_3; $[\overline{O_3}]$ is the average O_3 concentration during the time of observation.

If the rate of oxidation of M by O_3 is to be predicted, both $[\overline{O_3}]$ and k_M must be known. A few k_M values are given in Table 1, and many others have been published.

Equation (1) allows calculation of the ozonization time required to reduce an initial concentration of M to a given value by the direct reaction of ozone. For example, the time constant τ_M for the 'elimination' of M to 37 % (factor of e^{-1}) of its original concentration is:

$$\tau_M = \frac{1}{k_M[\overline{O_3}]} \qquad (2)$$

Such time constants, based on an average ozone concentration of about 0.5 mg/l (i.e. 1×10^{-5} molar), are included in Table 1. The entries in Table 1 show that the reactions between ozone and various solutes depend greatly on the nature of the compounds. For example, phenol at pH 8 (where it is approximately 1 % dissociated) has an apparent τ_M value of 1×10^{-2} s, whereas benzene has a τ_M of 5×10^4 s. In general, deprotonated phenols or ionized organic acids and neutral, basic nitrogen compounds (e.g. amines) react much faster than the corresponding protonated species. For this reason, the rate of some constants reported in Table 1 is given for pH 7 and 8.

Table 1. Second-order rate constants for the direct reaction of ozone with dissolved solutes and the time-constants for elimination of these solutes to 37 % of their original concentration [Data from Ref. 1, 2]

	k_{O_3} ($M^{-1}s^{-1}$) for pH = 7/(8)	τ_M(s)* for pH = 8
Ethanol	0.02	5.10^6
Cyclopentanol	2	5.10^4
Benzene	2	5.10^4
o-Xylene	10^2	10^3
Phenol	$10^6/(10^7)$	10^{-2}
3-Hexene-1-ol	$>10^5$	1
1,1-Dichloroethylene	10^2	10^3
Tetrachloroethylene	<0.1	$>10^6$
Ammonia	0.1/(1)	10^5
Glycine	$2.10^2/(2.10^3)$	50
Trimethylamine	$5.10^3/(5.10^4)$	2
Urea	<0.05	$>2.10^6$
Formate ion	10^2	10^3
Acetate ion	$<10^{-4}$	$>10^9$

* conditions and assumptions:

(1) $[\overline{O_3}] = 0.48$ mg/l ($= 10^{-5}$ M);

(2) one-to-one stoichiometry between ozone and solute;

(3) plug-flow or batch-type reactor.

OXIDATION BY OH˙ RADICALS

For example, about half of the ozone introduced into Lake Zurich water decomposes within 10 min, a typical time for an ozonization process. However, whenever ozone decomposes in water, about half of it becomes converted to OH˙ radicals, the most reactive oxidant known to occur in water. They even oxidize humic acids and carbonate or bicarbonate ions. Because of this high reactivity of OH˙ radicals, the oxidation of a specific compound M by OH˙ radicals will only occur in competition with the oxidation of all 'scavenger' materials, e.g., HCO_3^-, and humic acids, present. If $O_{3,\Delta}$ represents the amount of ozone decomposed to OH˙ radicals, we may therefore write:

$$O_{3,\Delta} \rightarrow OH^{\cdot} \begin{cases} \xrightarrow{+M} & OH^{\cdot} \text{ consumed by reaction with M} \\ \xrightarrow{+\Sigma S} & OH^{\cdot} \text{ consumed by reaction with scavengers S} \end{cases} \quad (3)$$

Assuming that M is a micropollutant which only consumes a small part of all OH˙ radicals reacting with S, the oxidation of M by OH˙ radicals formed from decomposed O_3 can be formulated by:

$$-\ln([M]_t/[M]_0) = O_{3,\Delta}/\Omega_M \quad (4)$$

where $[M]_t/[M]_0$ is the relative residual concentration of M after an amount of $O^0_{3,\Delta}$ has decomposed in the water. Ω_M can be called the *oxidation-competition value* of the water for the micropollutant M. Its value increases with the concentration of all OH˙ consuming scavengers present and decreases with the relative rate constant with which M reacts with OH˙ radicals:

$$\Omega_M \alpha \Sigma\,(k_i'[S_i])/k_M \quad (5)$$

There k_i' and k_M' are the second-order rate constants for the reactions of OH˙ radicals with individual scavengers, S_i, and the specified micropollutant M . Ω is a composite value. It can be expressed as the sum of the effect of all the individual scavenger concentrations when these concentrations are multiplied by the appropriate individual *oxidation-competition coefficients*, ω_i:

$$\Omega_{M^{\cdot}} = \sum([S_i]\omega_{M,i}) \quad (6)$$

These coefficients, examples of which are presented in Table 2, can be determined easily for many constituents of natural waters. Since the nature of solutes in natural waters is not well defined, a direct calibration of the waters by a measurement of a Ω value is recommended. Such measurements have been made for a number of different waters in Switzerland. The Ω′ values reported in Figure 1 are for the reduction of the concentration of benzene to 37 % of its original value. For easier comparison the Ω values were corrected to Ω′ values by subtracting the contribution of bicarbonate. This correction was always less than 0.3 g $O_{3,\Delta}/m^3$. The Ω′ values are given in Figure 1 as a function of the UV absorbance at 255 nm

Table 2. Oxidation competition coefficients, $\omega_{M,i}$ of different types of dissolved materials acting as $OH^{\cdot}$ radical scavengers (S_i)

S_i	$\omega_{M,i}$(g/g)
HCO_3^-/CO_3^{2-}	0.0016/0.05
NH_4^+/NH_3	0.000/0.12
H_4SiO_4	0.000
Cl^-	<0.003
1-Octanol	0.5
Glucose	0.13
DOC of natural surface waters	0.25
Fulvic acid (extracted from soil)	0.2

which was linearly correlated with the DOC also indicated on the abscissa. The plot reveals that Ω' increases by 0.25 g $O_{3,\Delta}/m^3$ for each additional g DOC/m^3.

This means that the ozone required to remove a specific compound (pollutant) may be appreciably increased by high concentrations of dissolved organic

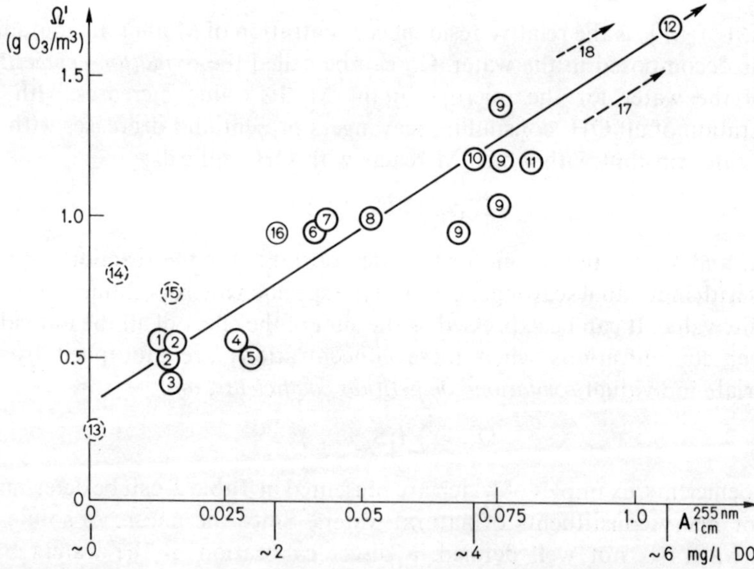

Figure 1. Oxidation-competition values of different types of water, based on benzene as reference solute. Ω' is corrected for effects of bicarbonate. Water samples from: 1 Lake Lucerne, 2 Lake Geneva, 3 Lake Lugano, 4 Lake Zurich, 5 Lake Constance, 6 Lake Biel, 7 Rhine River above Basel, 8 Small mountain lake, 9 Greifensee, 10 Lac de Morat, 11 Lac de Bret, 12 Lützelsee, 13 Mountain ground water (Engadin), 14 Well in a woods, 15 Groundwater, Schaffhausen, 16 Groundwater, Dübendorf, 17 Diluted secondary sewage effluent before nitrification, 18 Diluted secondary sewage effluent after nitrification

material. Pre-filtration or prechlorination hardly changed the Ω' values of these waters. The rate of reaction of a solute M_2 can be estimated from the rate of removal of a reference solute M_1 if their relative rate constants, k'/k'_2, for the reactions with $OH^{\cdot}$ radicals are known:

$$\Omega_1/\Omega_2 - k'_2/k'_1 \quad (7)$$

This formula can also be deduced from equation (5). Many rate constants k'_i for $OH^{\cdot}$ radical reactions are given in the literature reported.

SOLUTE-INDUCED OXIDATION REACTIONS

Intensive ozonization of water which contains significant concentrations of bromide (e.g., seawater) or manganous ion (such as may be present in anaerobic groundwater) may result in the formation of hypobromite or permanganate ions. These species may react as secondary oxidants or disinfectants. The effects of such secondary oxidants may depend on the solutes present and the characteristics of the ozonization process. They must be evaluated experimentally whenever significant concentrations of bromide or manganous ions are present.

NECESSITY OF DISTINGUISHING THE DIFFERENT OXIDATION PATHWAYS

Extrapolations of the extent and products of ozonization reactions from one water type to another can be valid only if the different reaction pathways are distinguished. The extent of the direct reaction increases with time and average concentration of molecular ozone present, but the extent of $OH^{\cdot}$ radical reactions increases with the amount of ozone decomposed to $OH^{\cdot}$ radicals and decreases with increasing concentrations of scavengers which react with $OH^{\cdot}$ radicals.

Of further importance to distinguish the different pathways of ozonization is the fact that they may lead to different types of products.

However, a separation of the contributions of these different reaction pathways is generally easily made; the direct reaction of ozone is so selective that, based on lists of reaction-rate constants, a clear distinction can be made between compounds which can be oxidized by that route and those which are not significantly oxidized during the ozonization process. For the latter compounds oxidation by $OH^{\cdot}$ radicals can be tested to be the main reaction of significance if their efficiency is depressed as expected by the addition of further radical scavengers.

Because of the high selectivity of direct ozonization reactions, oxidation products are observed to accumulate which are themselves only relatively slowly attacked further by ozone. For example, ozonization of natural surface waters results in the rapid production of aldehydes by the ozonolysis of olefinic compounds. Then, if ozonization is continued, short-chain carboxylic acids accumulate. These, in turn, are only very slowly oxidized, even by reactions involving $OH^{\cdot}$ radicals.

FURTHER READING

J. Hoigné, Mechanisms, rates and selectivities of oxidations of organic compounds initiated by ozonation of water. In: *Ozone Techn. and its Practical Application*, R. G. Rice and A. Netzer (eds) (Ann Arbor 1981; in press)

J. Hoigné and H. Bader, *ACS Symp. Ser.*, **82,** 292 (1978)

J. Hoigné and H. Bader, Ozone. *Sci. and Eng.*, **1** (2), 357–72 (1979)

Farhataziz and A. B. Ross, *Selected Specific Rates of Reactions of Transients from Water in Aqueous Solution. III: Hydroxyl Radical and Perhydroxyl Radical and their Radical Ion.* National Bureau of Standards, NSRDS-NBS 59 (Washington, DC: US Government Printing Office, 1972)

J. Hoigné and H. Bader, Kinetic und Selektivität der Ozonung organischer Stoffe in Trinkwasser. In *Wasser Berlin '77, Int. Symp. Ozon und Wasser* (Colloqium Verlag AMK, Berlin, 1978)

CHAPTER 12

Action of ozone on humic compounds

D. MAIER

Betriebshabil- und Forschungslabor der Bodensee-Wasserversorgung

INTRODUCTION

This contribution is a short summary of a large research work carried out by both literature research and practical investigations on the action of ozone on humic substances dissolved in the water of Lake Constance. The diversity of the humic compounds in the demonstrated structures incites us to attempt a general description of the action of ozone that could be applicable to all molecular structures.

The consequences of the general action of ozone in the different treatment stages and the specific action of ozone will both be discussed within the scope of this manual. These aspects are detailed by the twelve general guidelines given hereafter which are to be particularly considered where the ozonization of water containing humic substances is concerned.

GUIDELINES

(1) The action of ozone on natural organic substances dissolved in water results principally in a change in molecular constitution. By ozonization the polarity of the molecules present is increased and as a consequence they are more hydrophilic and contain more oxygen atoms in their structure, therefore less ethylenic double bonds. Moreover the remaining molecules have a minor molecular weight and acid properties as they contain more carboxyl groups which are easier to carboxylate, oxidize or decompose biologically. In some instances the formed molecules can have properties promoting flocculation. On the other hand the adsorption equilibrium on activated carbon is less favourable, but the formation capacity of chlorinated compounds is lower.

(2) In the case of dissolved humic compounds no significant oxidation to CO_2 is observed.

(3) In post-ozonization, as a refining technique, the use of ozone is ideal to destroy natural dyestuffs and compounds releasing taste and odour. 1 g of ozone per g dissolved organic carbon is the maximum dosing rate applicable to natural humic substances.

(4) In practice, the improvement of biodegradability by bacterial consumption of ozonized compounds results in a distinctly higher capacity of aftergrowth where ozonized waters are concerned in comparison to native waters especially in the case of reinfection.

(5) Ozonized water must contain a safety disinfectant for its transport in the distribution mains.

(6) When chlorine is used for post-disinfection the ozonization enables one to maintain a better chlorine residual as well as to obtain a 50% reduction of the chloroform concentration. The eventual presence of bromide in water must be considered carefully. In such a case one is likely to find more organo-bromine compounds after ozonization.

(7) Calcium salts can be precipitated by the modified humic acids formed through the treatment of the natural compounds with ozone. Also some organic impurities can be destabilized. Consequently, the ozonization stage is best installed prior to a final filtration.

(8) In the combined ozone-coagulation process the ozone is preferably put in as the first stage. In some instances the flocculation-aid properties of the oxidized humic compounds enable one to reduce the needs of flocculants and as a consequence, the costs. Multilayer filtration after flocculation is more and more used in this respect.

(9) Ozonization can also be considered as a technique to be placed in the middle of the treatment scheme's, i.e. to place the mechanical treatment processes before the ozonization step as these processes enable a significant decrease of the ozone needs.

(10) In many waterworks an activated carbon treatment is applied after the ozonization stage, in spite of the lowered adsorption rates of the organic compounds after ozonization. However, ozonization can be applied as a reliable demanganization method when followed by filtration on activated carbon. Furthermore, the biological purification on activated carbon is often enhanced by ozonization.

(11) Most of the effects of the treatment on the naturally occurring organic compounds are already obtained when using 0.5–1 g O_3 per g total organic carbon. The results are often less favourable when higher ozone levels are used and thus as a result of too high residual ozone concentrations as, for

example, after a too short contact time or owing to the disagreeable odour of ozone.

(12) In each particular case in which an ozonization stage is to be integrated in a pre-existing treatment scheme a preliminary investigation on the possible specific consequences of the ozone use must be undertaken. In this phase the necessary fundamental chemical research is to be achieved.

CHAPTER 13

Removal of bacteria and viruses by ozonization

J. C. BLOCK
UER d'Ecologie, Metz

Used for the disinfection of drinking water as early as 1906 in Nice (France), ozone has since been the subject of numerous studies. The aims of treatment differ according as to whether ozone is applied to drinking water or wastewater. In the first case, disinfection in the strict sense is required, in other terms the elimination of pathogenic micro-organisms and indicators of faecal pollution (*Escherichia coli* and *Streptoccus faecalis*). In the second case, a reduction in pathogenic micro-organic populations and indicators is sought.

Various parameters have an effect on the efficiency of treatment. Those considered most often in the literature are pH, temperature, organic matter, suspended solids, the concentration in dosable dissolved ozone (or ozone residual), and ozonized air–water contacting systems.

pH

At alkaline pH, ozone decomposes more rapidly. If, however, a constant residual is maintained, the bactericidal and viricidal activity remains stable in the pH range 5.6 to 9.8.

TEMPERATURE

Variations in temperature in the 0 to 37°C range have relatively little effect on ozone's disinfectant efficiency. In the most extreme cases, a rise in temperature increases the kinetic inactivation of micro-organisms, in agreement with the theory.

ORGANICS

Organic matter in water gives rise to a considerable oxidant demand whereby the appearance of a dissolved ozone residual in water takes place only when the applied dosage surpasses a critical amount corresponding to the ozone demand.

Prior to the appearance of such a residual, an oxidation of the organic matter can be observed and, at the same time, an inactivation of microorganisms up to 1 or 2 logarithmic decrease rates.

This highly interesting phenomenon for wastewater treatment might be explained by the existence of a zone of high ozone concentration at the interface of the ozonized air–water phase; which has led to proposals for new turbines for improved ozone dispersion or to special study of injection systems.

SUSPENDED SOLIDS

Suspended solids act on many occasions as a vehicle for the micro-organisms which are adsorbed on them. In most cases this adsorption limits the oxidant's effect and diminishes the extent of disinfection, as has been demonstrated for poliomyelitic viruses adsorbed on kaolin. Similar observations are moreover often made during ozonization of wastewaters for which a fraction of micro-organic populations are hard to inactivate, even at high dose rates.

RESIDUAL OZONE

The presence of residual ozone is generally considered a key parameter in the completion of water disinfection.

When an ozone residual appears at an injected critical dose level, the curves of micro-organism survival present a marked discontinuity, representing an increase in the kinetic inactivation of germs, which has been called an all-or-nothing effect. In the presence of an ozone residual at concentrations near 0.1 to 0.4 mg/l, excepting limitations due to aggregates or to adsorptions of micro-organisms, the kinetic inactivation of micro-organisms, in particular of viruses, can be described by a reaction of the first order. This inactivation results from a sizeable increase in the oxidation–reduction potential, and takes place relatively quickly, in 10 s to 5 min, depending on the micro-organisms.

The mechanisms of micro-organism inactivation (bacteria, viruses, protozoa) are actually more or less well known.

It appears that the action of ozone (or of free radicals formed in water by its decomposition) on bacteria affects the constituents of cytoplasmic membranes, enzymatic systems, and even nucleic acids; that lysis of the bacterial cell occurs only at high ozone concentrations; and for viruses, that ozone attacks the proteins of the capside to the same extent as the nucleic acids.

The sensitivity of these micro-organisms to ozone has been tested extensively in the course of comparative studies. Several observations can be made:

(a) faecal bacteria often appear more sensitive than enteroviruses;
(b) a number of contradictions can be noted among the conclusions of the various authors, which may be explained by differing work methods and by

the fact that curves for micro-organism survival can intercross and thus modify the relative sensitivity scale according to the level of inactivation considered;

(c) study of the inactivation of endogenous faecal bacteria and of endogenous viruses in wastewaters makes a greater differentiation in the sensibility to ozone appear in these different groups; viruses being far more resistant.

If one considers the rapidity of ozone's action due to its oxidation–reduction potential, its efficiency in water treatment, and the low toxicity it gives rise to in treated effluents, ozonization can be considered an excellent method of water disinfection.

FURTHER READING

J. C. Block, Contribution à l'étude de la désinfection des eaux d'alimentation par l'ozone. Thesis (1977)

S. Farooq, E. S. K. Chian and R. S. Engelbrecht, Basic concepts in disinfection with ozone. *J. Water Pollution Control Fed.*, **49,** 1818–31 (1977)

H. I. Shuval and E. Katzenelson, *Detection and Inactivation of enteric Viruses in Wastewater.* Environmental Protection Technol. Series, E.P.A. 600/2.77.095 (1977)

CHAPTER 14

Effect of ozone on the elimination of various algae by filtration

J. C. GINOCCHIO

Sulzer Brothers Ltd

The effects of preozonization before filtration or flocculation–filtration of lake waters are generally favourable to increase the elimination rates. The elimination of *Fragilaria crotonensis*, *Synedra*, *Tabellaria fenestrata*, *Cyclotella*, *Phacotus lenticularis*, *Asterionella formosa*, *Stephanodiscus* and *Oscillatoria rubescens* provides for suitable criteria. During the investigation the counts of the raw water varied by a factor of 100, e.g. from 10^3 to 10^5 per litre for *Stephanodiscus*.

Preliminary ozonization took place in a contact column. The ozone input amounted to about 2 mg/l and the reaction time 10 min. The natural ozone consumption induced by the reducing substances present in the water characterized the lake water tested. Figure 1 shows this consumption as a function of different initial ozone concentrations in the lake water. Curve 3

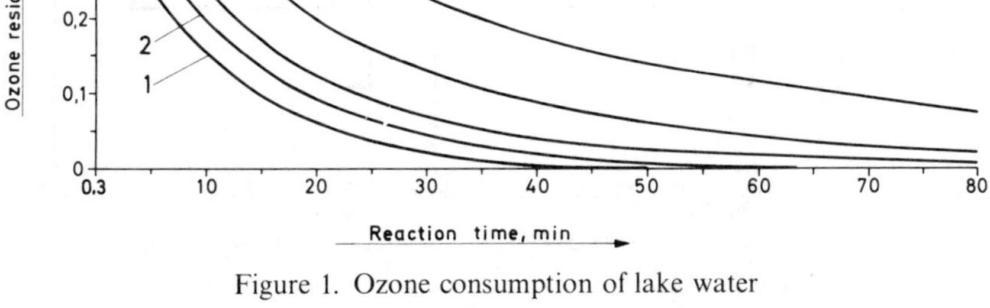

Figure 1. Ozone consumption of lake water

corresponds to our working conditions, i.e. at an ozone input of 2 mg/l, a residual ozone concentration of about 0.25 mg/l was found after 10 min contact time.

Figure 2 shows the results of the filtration without preozonization (a) through the coarse sand (1–1.5 mm size) filter. It can be seen from this diagram that the small plankton algae, such as *Stephanodiscus*, pass practically completely through this filter media. Information concerning the sizes and shapes of the plankton algae is given in this and the following figures. Preliminary ozonization (b) before coarse filtration has a positive influence on algae removal.

Figure 3(a) is a graphical representation of the elimination percentages achieved using fine sand (0.5–1.2 mm) filtration. The lowest filtration performance is obtained with respect to *Synedra*, *Cyclotella*, *Stephanodiscus* and *Oscillatoria rubescens*. Using fine sand filtration with preliminary ozonization a

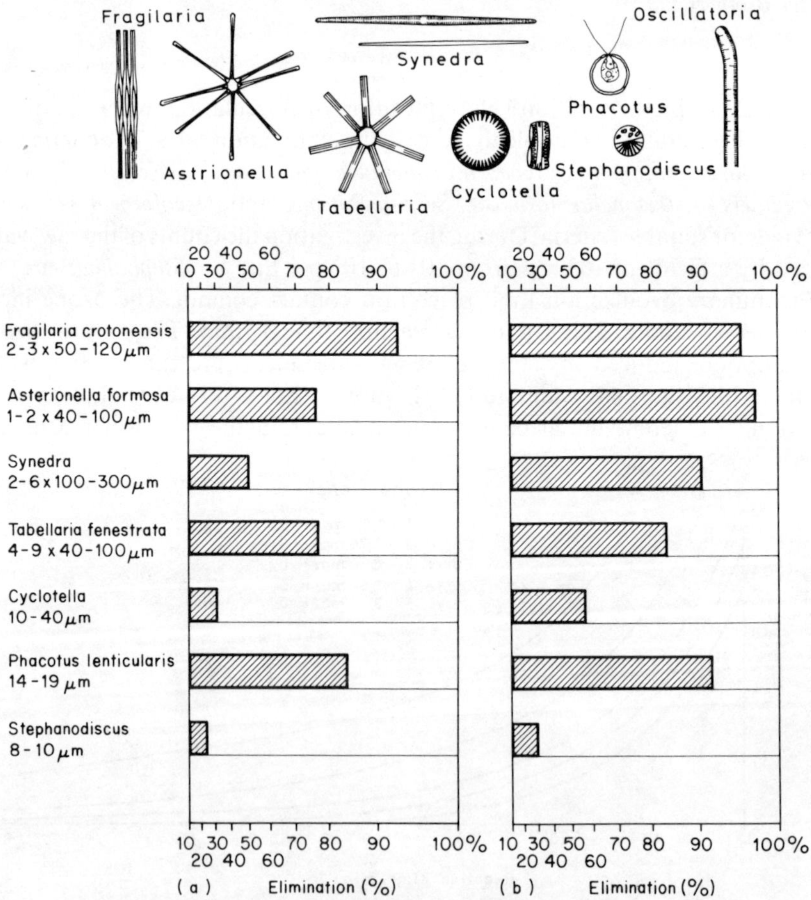

Figure 2. Ozonization–sand filtration for algae removal

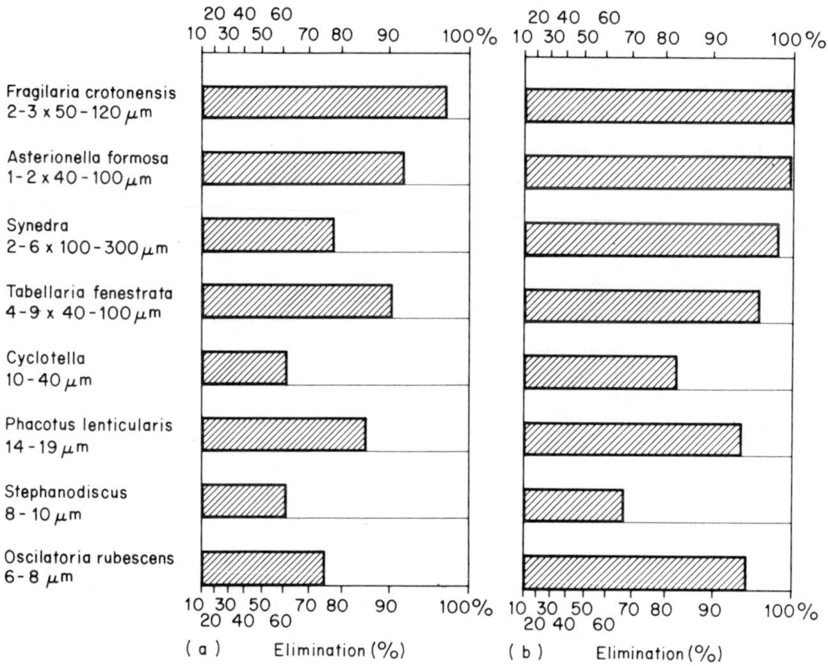

Figure 3. Ozonization–fine sand filtration

third of the *Stephanodiscus* passes through the filter, but the retention of *Oscillatoria* is good (Figure 3(b)).

The performance of flocculation–filtration was carried out using 2 mg/l aluminium sulphate and the double-layer filter (300 mm pumice of 1.5–2 mm grain size on a 600 mm sand layer 0.5–1.2 mm). However, the best removal of algae is obtained using flocculation–filtration in the two-layer filter with preliminary ozonization (Figure 4(b) compared to 4(a)).

Preozonization before filtration has always been observed to have a positive effect in the removal of algae.

The following is observed with regard to the effect of ozone on these plankton algae. *Asterionella*, *Synedra*, *Tabellaria* and *Cyclotella* were not affected. The sole exception among the diatoms was *Fragilaria crotonensis*, whose chromatophores largely lost their colour. *Oscillatoria rubescens* appeared to be somewhat more strongly entwined, possibly due to the loss of gelatinous material.

It can be stated that flocculation–filtration in the two-layer filter with preliminary ozonization is an excellent process, not only with respect to the high degree of algae elimination obtained, but also with regard to filter operating times. Hence high water quality is coupled with economic operation. Even filter layers of small thickness can be very efficient in this technique.

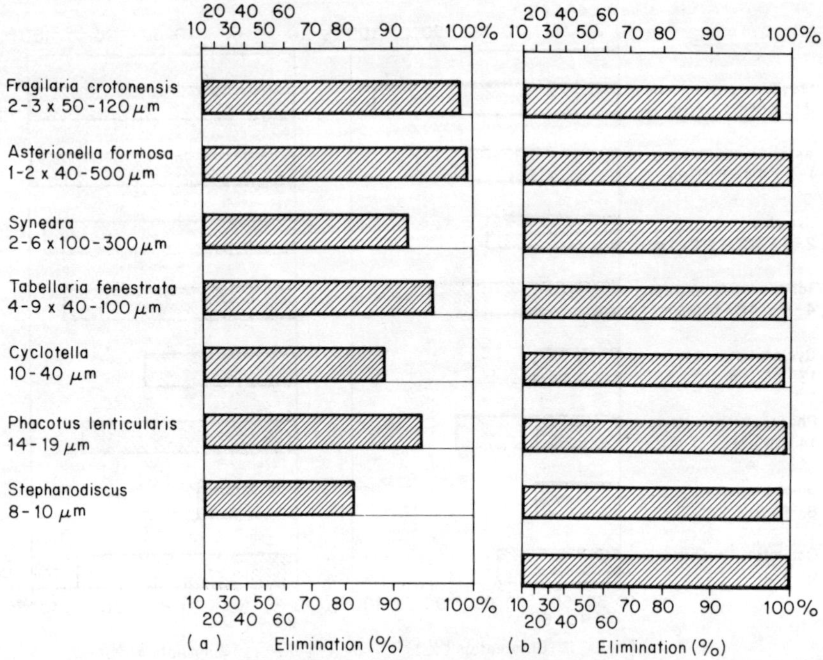

Figure 4. Preozonization–double layer filtration

FURTHER READING

U. Zimmerman, *Vergleichende Planktonuntersuchungen im Zürich-, Zürichober- und Walensee 1974 bis 1979*. Arbeitsgemeinschaft Wasserwerke Bodensee–Rhein AWBR. Annual Report (1979).

A. Wetzel, *Technische Hydrobiologie*. Akademische Verlagsgesellschaft Geest & Portig K.-G. Leipzig (1969).

CHAPTER 15

Action of ozone on various organic compounds

A. LAPLANCHE and G. MARTIN

CEEA–ENSC, Rennes

The smallest organic molecules often remain in solution after conventional treatment and can be attacked through oxidation e.g. with ozone. The following should be considered with special interest:

(a) alcohols, particularly ethyl alcohol which could be present after biological denitrification;
(b) aliphatic amines involved in most problems related to odour;
(c) Pesticides, as more or less permanent toxic residuals;

It is interesting to identify the oxidation products formed. It is necessary to establish that such compounds (amines, alcohols) or the more harmful (pesticides) do not produce, by the ozone treatment, molecules more toxic than the original products.

ALCOHOLS

Generally, the ozonization of alcohols follows the oxidation pathway shown in Formula 1, according to Bailey.[1] However, in a concentrated solution numerous

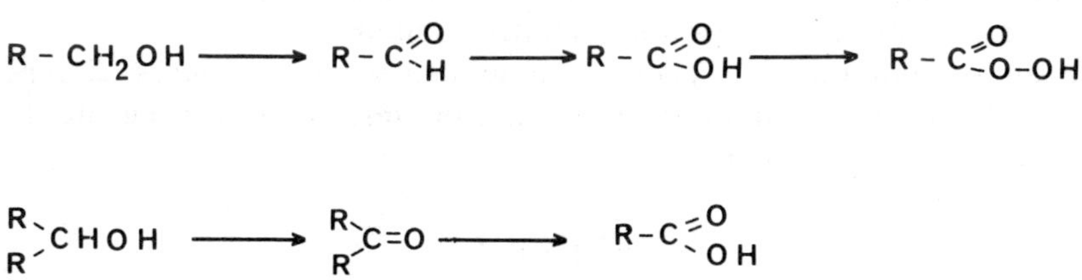

Formula 1

other products of oxidation are also formed. Thus for ethyl alcohol, Faujour[2] has obtained besides the acetaldehyde and acetic acid expected:

(a) Peroxidic compounds: H_2O_2 and organic peroxides. The formula has been attributed to one of the two

$$CH_3-\underset{\underset{O}{\|}}{C}-O-\underset{\underset{OOH}{|}}{CH}-CH_3$$

(b) Condensation products: compounds similar to polyethylene glycols besides ethyl acetate, acetaldehyde, diethyl acetal.

For diluted solution of ethyl alcohol (8 mg/l) even after 2 hours of ozonization, only acetaldehyde and acetic acid are formed. The peroxides are not detected. Therefore ozonization is feasible without significant risks, particularly if only traces of ethyl alcohol are present.

For isopropyl alcohol, only acetone formation is mentioned by Kuo,[3] whereas Dalouche *et al.*[4] have identified acetone, acetic acid and isopropyl acetate. In a *concentrated* solution they indicate the formation of peroxidic compounds. In both cases the mechanism begins by an electrophilic attack of ozone on oxygen.

ALIPHATIC AMINES

In concentrated solution, the compounds resulting from the ozonization of different classes of amines are divided into four different groups according to the reaction mechanisms involved:

(a) fixation of an atom of oxygen on a nitrogen: hydroxylamine, oxime, aminoxide;
(b) fixation of an atom of oxygen on carbon in α of nitrogen: amide results;
(c) breakdown of the $-\overset{\diagdown}{\underset{\diagup}{C}}-N\overset{\diagup}{\diagdown}$ bond: mineral nitrogen (NO_2^-, NO_3^-), aldehydes, carboxylic acids;
(d) secondary condensation reactions of polyamide types.

The mechanism starts with an electrophilic attack of ozone on the nitrogen atom,[1,4] followed either by oxidation of the carbon chain in α of nitrogen or by a general breakdown of the molecule. Finally condensation products are formed. They are however not observed in a diluted solution.

In an ozonization pilot plant, in a diluted solution, amines disappear completely in a few minutes. It seems therefore interesting to eliminate these compounds by ozonization.

PESTICIDES

Organochlorous pesticides are not easily oxidized and need exhaustive ozonization conditions. Specific reactions are the epoxidation of aldrin and heptachlor in dieldrin and heptachlorepoxy (these compounds are easily adsorbed on non-saturated active carbon). Organophosphorous pesticides are very sensitive to ozonization. Numerous products have been studied: parathion, malathion, phosalone, diethyldithiophosphoric acid, dimefox, etc.

In these cases, several products are obtained:

(a) oxidation of P=S bond with formation of oxons (P=O);
(b) degradation delivering H_3PO_4 or condensation products of the phosphor part;
(c) degradation or condensation products of phosphorous substitutes;
(d) oxidation of carbon chains in α of nitrogen in phosphorous amidate compounds.

In the case of parathion, in a very diluted solution (80 mg/m^3) the kinetics of the initial oxidation is as follows:

$$\frac{-d(\text{parathion})}{dt} = K\ [\text{parathion}]\ [\text{dissolved ozone}] \qquad (1)$$

with $K = 4.20 . 10^3\ \text{mn}^{-1}$ for pH = 7.5 and $t = 20°C$.

This means a half-life varying from 17 to 2 min with concentrations of ozone varying from 0.45 to 3.30 mg/l of water. For similar concentrations, one obtains a total disappearance in a few minutes of parathion methyl, malathion, fenitrothion and ronnel.

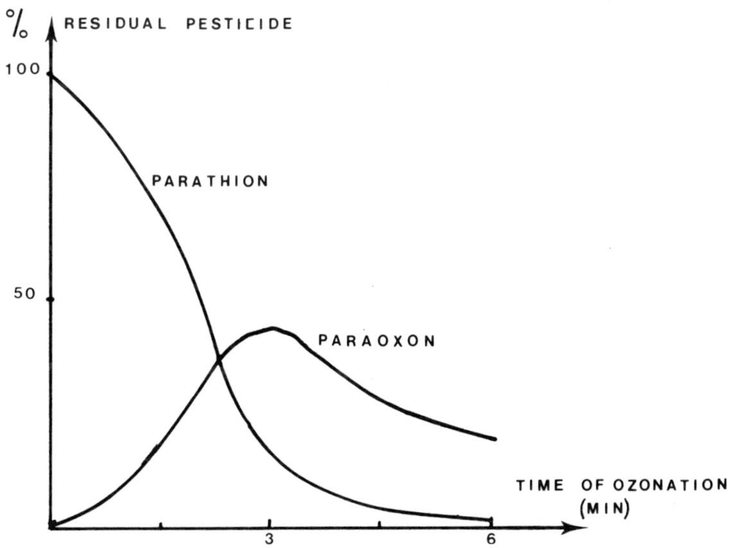

Figure 1. Ozonation pilot plant (ozonization of parathion) O_3: 3 mg/l of water

Other parameters intervening in the reactions are, besides pH and temperature, the physical characteristics of the distribution of oxidizing agent. The available reaction time for a given dose can have a determining effect on the success of the process.

REFERENCES

1. P. S. Bailey, *Ozone in Water and Waste Water Treatment*, F. L. Evans (ed.), Ann Arbor Science Publication (1972).
2. M. Elmghari-Tabib, A. Dalouche, C. Faujour, F. Venien and G. Martin, *Proceedings of 5th International Congress 10A, Berlin*, Colloquium-Verlag, O. H. Hess, Berlin (1981).
3. P. P. K. Kuo, E. S. K. Chian and B. J. Chang, *Environ. Sci. Techn.*, **11** (13), 1177 (1977)
4. A. Dalouche, F. Venien and G. Martin, *Environ. Techn. Let.*, 1981. (In preparation)

CHAPTER 16

Ozone and chlorinated organic compounds

M. Doré

UER Sciences fondamentales et appliquées, Poitiers

The chlorination of water gives rise, by its action on organic micropollutants, to the formation of trihalomethanes. The compounds formed at the different stages of chlorination (prechlorination or disinfection) are essentially chloroform ($CHCl_3$), bromodichloromethane ($CHBrCl_2$), dibromochloromethane ($CHBr_2Cl$) and bromoform ($CHBr_3$).

When using ozone and chlorine in the same flow sheet for treatment, the incidence of ozonization on the formation of these compounds must be considered, particularly where the action of ozone on the precursors (direct action) and on the intermediates of the haloform reaction (indirect action) is concerned.

PRECURSORS OF THE HALOFORM REACTION

Fundamentally, the formation of trihalomethanes results from the substitution by halogens of the mobile hydrogen atoms carried by a carbon atom, followed by the cleavage of the molecule (see Formula 1). Among the most reactive precursor compounds, the humic acids, the polyphenols and the β-diketones must be mentioned.

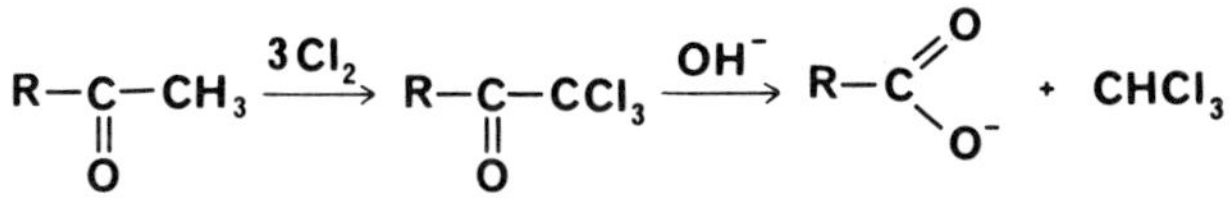

Formula 1

DIRECT ACTION OF OZONE ON THE PRECURSORS

Because of its high oxidizing power ($E_0 = 2.07$ V), ozone is potentially capable of giving rise to the formation of precursor compounds, by acting on molecules which are initially inert to chlorine (P.THM = trihalomethane formation potential) (see Formula 2). Indeed, the increase in trihalomethanes by chlorination of aqueous solutions of preozonized phenols at low rates of treatment is observable, whereas the preozonization of resorcinol leads to a

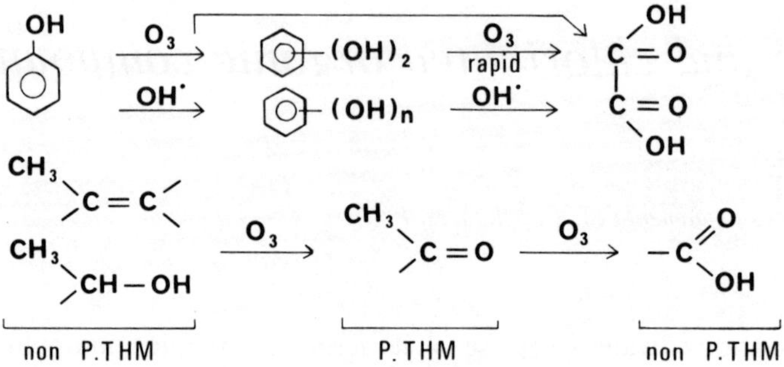

Formula 2

decrease in the quantity of volatile organochlorinated compounds. This decrease is most important as the rate of ozonization used is greater. In practice, in most cases, the effect of ozonization is to reduce the concentration in precursors, as shown by the curves in Figure 2. It must be noticed however that some results show a contrary effect.

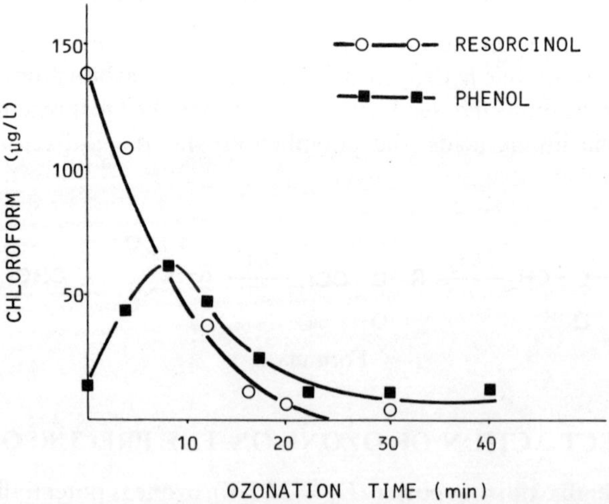

Figure 1. Action of ozone on phenol and resorcinol related to the THM.FP

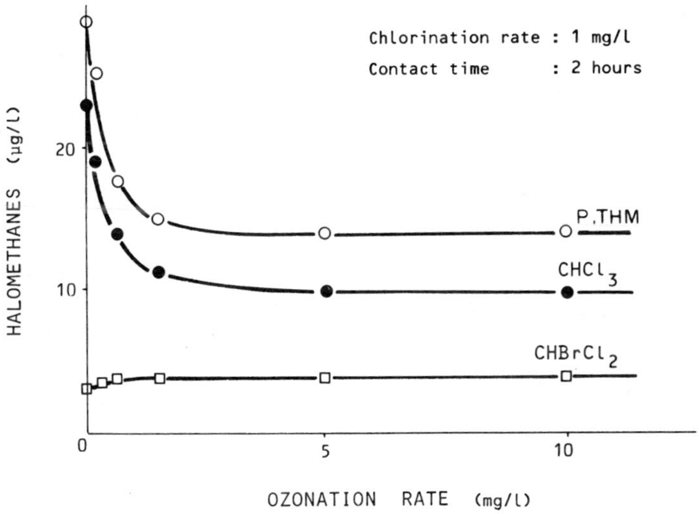

Figure 2. General trends of ozonization *vs.* THM.FP

INDIRECT ACTION OF OZONE ON THE HALOFORM REACTION

Oxidation of halides

Taking into account the respective values of the redox potentials of the couples O_3/O_2, $HOCl/Cl^-$ and $HOBr/Br^-$ (2.07 V, 1.49 V and 1.33 V), chlorine and ozone can oxidize the bromides contained in water. The hypobromous acid thus formed could then, by acting on the precursors, lead to the formation of bromoform and compete with the chlorine to form mixed compounds, according to a pathway of the type indicated in Formula 3. As generally observed ozone has no notable oxidizing action on chlorides and therefore cannot produce hypochlorous acid by acting on the chlorides contained in water.

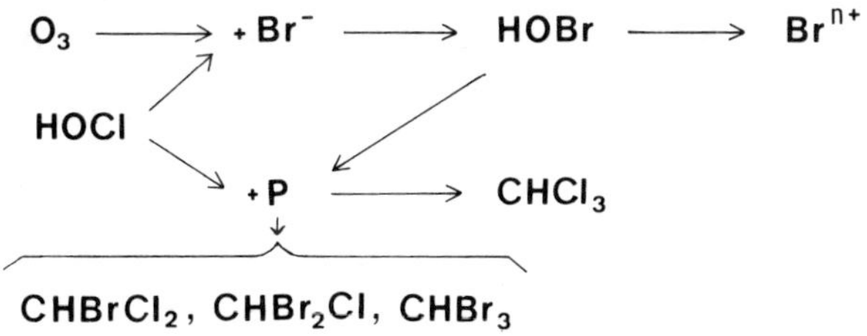

Formula 3

In practice, the influence of ozonization on the nature of volatile organohalogenated compounds has not been fully clarified. Indeed the formation of bromoform by ozonization of sea water has been observed whereas other results seem to show that ozonization has no notable influence. These last results may perhaps be explained by a more exhaustive oxidation of the bromides, for example into bromates, in certain operating conditions.

Role of ammonia

Ammonia, by its rapid action on chlorine, to produce chloramines, competes with the precursors. However, taking into account the very high reactivity of certain precursors, ammonia cannot totally inhibit the formation of trihalomethanes (even before the break point).

As ozone has only a weak oxidizing power on the NH_4^+ ions, under normal ozonization conditions (pH close to neutral), it follows that a preozonization will have no direct incidence on the NH_4^+–precursors competition, in post-chlorination.

CONCLUSION

Although ozonization can potentially give rise to precursors of haloform formation, it most generally leads to the degradation of these compounds and to a reduction in the quantity of the trihalomethanes formed on post-chlorination.

Ozone has no significant oxidizing action on ammonia or chlorides, and its action on bromides within the range of concentrations usually found in fresh water, seems to be of limited extent. The formation of organobrominated compounds which takes place is usually very limited.

FURTHER READING

J. J. Rook, *Water Treat. Exam.*, **23,** 234–43 (1974)

Committee Report of the A.W.W.A., *J.A.W.W.A.*, **66,** 682–8 (1974)

E. A. Crecelius, *Symposium on Adv. Ozone Techn.*, Toronto, Int. Ozone Inst., Cleveland, Ohio (1977)

M. Doré, B. Langlais and B. Legube, *Water Research*, **12,** 413–25 (1978)

M. Doré, N. Merlet, T. Blanchard and B. Langlais, *Prog. Wat. Tech.*, **10,** 853–65 (1978)

M. D. Umphres *et al.*, *Ozonews*, **6** (3), (1979)

M. Doré, N. Merlet and J. De Laat, *Journal Français d'Hydrologie*, **10** (29), 131–40 (1979)

J. Hoigné, *Prog. Wat. Treatm.*, **10,** 1115 (1978)

CHAPTER 17

Principles of repeated ozonization

M. RAPINAT

Compagnie Générale des Eaux

Currently, the roles assigned to ozone in water treatment are multiple. They include:

(a) bacterial and viral disinfection;
(b) the precipitation of metals;
(c) flocculant action;
(d) taste, odour, and colour control;
(e) the degradation of organic substances.

If raw water is slightly polluted, only one of ozone's actions need be turned to, and application can be effected at a single point. In the case of heavily polluted water, it is worth while to employ various effects of ozonization. The problem then arises of how to integrate ozonization into a scheme of treatment composed of several process stages.

Studies of the mechanisms of the reactions of ozone with various pollutants indicate that it is often worth while to enhance a treatment stage by combining it with an ozonization.

The principal combinations that can be considered are the following:

PREOZONIZATION OF RAW WATER BEFORE STORAGE

The role of ozonization is to improve the natural phenomena of decantation and the elimination of organic matter that take place during storage. Ozone can also curtail the undesirable effects of several inhibiting metals, such as lead and copper. Here, the contact time can be brief and the dosage weak, e.g. in the order of 1 mg/l.

OZONIZATION PRIOR TO CLARIFICATION

With ozone, flocculant dosages can be reduced while the quality of the filtered water is maintained. In most cases there is an optimum rate above which ozone does not improve this phase of the treatment. This rate is near 1 mg/l. It is low enough so as not to create undesirable by-products from organic matter. Ozonization coupled with filtration proves very efficient, e.g., also for groundwaters containing iron and manganese.

OZONIZATION AND FILTRATION ON GRANULAR ACTIVATED CARBON

To improve treatment in plants faced with organic pollutants, it appears more and more indispensable to turn to filtration on granular activated carbon using biological processes.

Ozonization before filtration on carbon can increase the elimination by biological means of some of the organic substances. Further, for viricidal ozonization after clarification requiring a contact time of 10 min and heavy doses, activated carbon filtration as a means of retaining by-products of ozonization appears favourable. An additional advantage of the combination of ozone and granular activated carbon is that a more stable water can be obtained at the final stage. Post-treatment chlorination will still be necessary in most cases for the protection in the distribution system.

Every type of raw water is a sample case requiring treatment processes adapted to the pollutants to be eliminated. In a general manner, where surface water needs considerable ozonization, it seems advantageous to spread application over several points to improve the efficiency of several treatment stages. This technique optimizes treatment conditions at the different stages and adds additional flexibility. This is most notably the case for new biological lines of treatment where ozone enables biological purification through further treatment-steps.

FURTHER READING

M. Schalekamp, *Gaz–Eaux–Eaux Usées*, **57**, 657–73 (1977)

G. W. Miller, R. G. Rice, C. M. Robson *et al.*, *E.P.A. Municipal Environmental Research Laboratory*, 585 pp. E.P.A.–600/2–78–147 (August 1978)

R. G. Rice, C. M. Robson, G. W. Miller *et al.*, *Procédés d'Oxydation Appliqués au traitement de l'eau potable*, Karlsruhe, 11–13 September 1978; W. Kühn and H. Sontheimer (eds.), pp. 442–76. Institut Engler-Bunte de l'Université de Karlsruhe, Karlsruhe (1979)

Y. Richard, *L'eau et l'industrie*, no. 31, pp. 35–42 (January 1979)

H. Lienhard and H. Sontheimer, *European Forum on Ozone Applications to Water Treatment*, Nice, 25–26 January 1979; International Ozone Association European Committee (ed.), pp. 345–70 (1979)

L. Divet, Techniqus et sciences municipales, *L'eau*, no. 5, pp. 309–14 (1979)

CHAPTER 18

Influence of ozonization on the structure of organic matter and its biodegradability within activated carbon

M. Bourbigot* and M. Dore†

*Compagnie Générale des Eaux and † UER Sciences fondamentales et appliquées, Poitiers

The influence of ozonization on the structure of molecules or of families of molecules dissolved in water has already formed the subject of numerous studies.

Conclusions are always more or less the same: ozone breaks down aromatic nuclei, cleaves double bonds, makes organic matter more biodegradable and more polar by creating carboxylic, aldehyde and ketonic compounds, etc.

In the study of natural waters, that is to say of very complex mixtures of many compounds of which the greatest part cannot be identified, competitive effects play a role and one cannot foresee exactly at which ozone doses the phenomena mentioned above will take place, nor to what extent.

In practice, user needs for industrial treatment lines which include, as is more and more often the case, ozonization followed by filtration through activated carbon, constantly face the operator with the following question: what ozone dosage should be used to obtain optimum results? This means in terms of treated water quality, into maximum removal of total organic carbon, great residual chlorine stability where chlorine is used as bacteriostatic, and maximum removal of biodegradable organics, as organics can entail bacterial aftergrowth in the distribution system. In operation the organic compounds present cannot be identified yet on a continuous basis and therefore the plant operation relies on general parameters like UV-absorption, TOC, chromatographic analysis and evaluation of bacterial activity.

In industrial applications, ozonization has a real yet limited effect (from 5 to 15%) on the reduction of the TOC content of water before activated carbon filtration.

Oxidation of aromatic compounds by ozone can readily be evidenced in operation by a reduction of UV adsorption from 270 to 254 nm; the ozone dosage required for minimal absorption is variable. It is a function of treated water quality, and one generally reaches a plateau corresponding to a reduction of the optical density of the order of 50 % (Figure 1).

Ozonization entails an increase in biodegradability, determination of the most appropriate dosages requiring study for each type of water. Removal of nutrients by bacteria within the activated carbon is thus favoured.

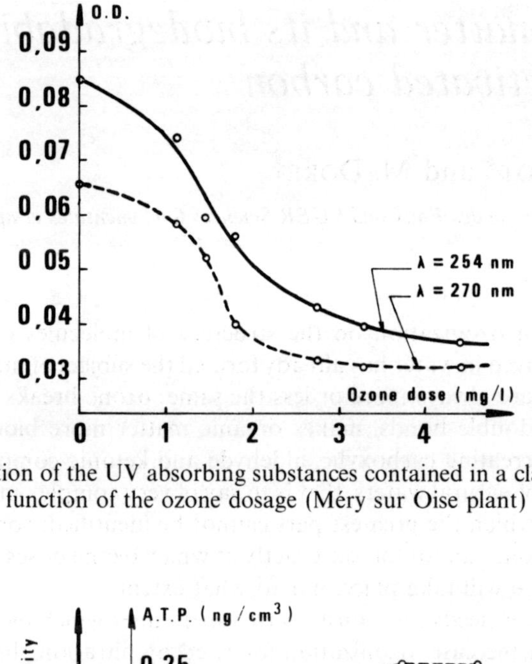

Figure 1. Reduction of the UV absorbing substances contained in a clarified water as a function of the ozone dosage (Méry sur Oise plant)

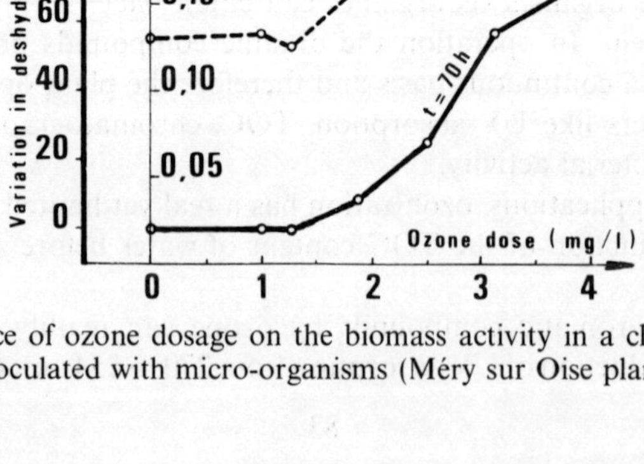

Figure 2. Influence of ozone dosage on the biomass activity in a clarified Oise water inoculated with micro-organisms (Méry sur Oise plant)

The positive effect on the biodegradability of organic matter in activated carbon largely offsets the negative effect on adsorption caused by increased polarity; further, the first is lasting while the effect on adsorption quickly fades. With optimum treatment a reduction of TOC between 20 and 40 % can be maintained over more than 2 years, depending on the quality of the water to be treated.

The effectiveness of the combination ozone + activated carbon can be hampered by two phenomena:

(a) insufficient contact time (excessive flow rates or inadequate carbon layer);
(b) a biodegradable compound load that is insufficient to assure adequate development of the zooglea and does not allow the biosorption process to build up and retain oxidized products that are not easily adsorbable on carbon.

Despite these restrictions, the effectiveness of the coupling ozone + activated carbon filtration is remarkable. This technique is used more and more frequently for treating waters loaded in organic matter because it is the only economically feasible process which can limit the content in nutrients for micro-organisms present in distribution systems. The drinking water treatment plant at Méry-sur-Oise can be cited as an example. Its treatment line limits consumption of residual chlorine used as bacteriostatic and thus makes possible the distribution of a water of good quality as to its flavour, its content in chlorinated organic compounds and its bacterial innocuousness.

FURTHER READING

B. Legube, B. Langlais, B. Sohm and M. Doré, p. 541 in *Proceedings of Wasser Berlin (1981)*, Colloquium Verlag, O. H. Hess (1981)

R. G. Rice, C. M. Robson, G. Wade Miller and A. G. Hill, Uses of ozone in drinking water treatment, *J.A.W.W.A.*, **73**, 44 (1981)

P. Schulhof, Water supply in the Paris suburbs: changing treatment for changing demands, *J.A.W.W.A.*, **72**, 428 (1980)

P. Schulhof, An evolutionary approach to activated carbon treatment, *J.A.W.W.A.*, **71**, 648 (1979)

R. Brunet, M. M. Bourbigot, B. Legube and M. Doré, Evolution Chimique des matières organiques au cours de l'ozonation d'une eau de surface filtrée, *Aqua*, **4**, 76–7 (1980) (13th IWSA Congress, Paris, 1–4 September, 1980)

R. Brunet, M. M. Bourbigot and M. Doré, p. 665 in *Proceedings of Wasser Berlin (1981)*, Colloquium Verlag, O. H. Hess (1981)

J. Hoigné, *Prog. Wat. Techn.*, **10**, 1115 (1978)

CHAPTER 19

Micellization and improvement of coagulation by ozonization

Y. RICHARD

Degrémont

Ozone is not, strictly speaking, a coagulant; it has no action on the electric charges of the colloidal matter of the water; however, in certain well-defined cases such as deep lake waters, a slight coagulant effect due to the ozone can be noticed.

Some of the impurities contained in these waters are heterogeneous complexes which bind the natural matter to iron and manganese. These waters may be coloured but their turbidity is often low. Ozone destroys the chromophores and reduces the colour of these waters. It liberates the iron and manganese from these organic complexes. If the required pH conditions are reached there results the formation of an often fragile, low-volume metal precipitate which can be retained by direct flocculation on a filter. If the flocculation by itself is fast enough the filtration can be made directly without the addition of any flocculant. However, if the flocculation takes place only with some difficulty the addition of a flocculant will be necessary.

In rarer cases, and in the same time as the decolorization of these waters, a small increase of the turbidity and the formation of a very light floc of organic origin can be noted. This effect is called micellization. The formation of this floc results from the new equilibrium between the organic solutes with a higher polarity and the water. Particle-counting and classification according to their size show, after the action of ozone, a change in the distribution curve of the molecules according to their size. For low treatment rates the number of large molecules increases, thus revealing a polymerization or flocculation of certain particles. Too high an increase of the treatment rate must be avoided as this leads to a destruction of the condensed particles and an increase of the small particles which are difficult to flocculate.

The addition of a small amount of coagulant easily neutralizes part of the colloidal charge and completes the action of the ozone. The amount of coagulant to be used varies between 2 and 5 g/m^3 of aluminium sulphate or ferric chloride; in order to improve the retention of the floc which is formed in the filtering material it is sometimes necessary to add a small amount of flocculation aid.

In the case of surface waters characterized by a higher content of suspended solids and of dissolved organic matter than the preceding waters, it is most often necessary to use a clarification treatment by coagulation, flocculation, settling and filtration. The introduction of ozone at the beginning of the treatment can promote, under certain circumstances, the condensation of certain molecules and thus lead to an improvement of the flocculation. Thus, a low amount of ozone may lead to a reduction of the treatment rate. This effect of the ozone is all the more important in that the amount of coagulant is lower than the optimal treatment rate (defined as the necessary treatment rate to cancel the Zeta potential of the colloids). Too large amounts of ozone, however, impede flocculation.

However, on other waters it must be noticed that pre-ozonization had only a very slight effect which could even be reduced to zero, e.g. when the water contains only a small amount of organic matter.

To conclude it may be stated that the use of ozone as a pretreatment process—as a coagulation technique—leads to interesting results although preliminary tests are generally necessary to determine the application conditions. Also the results may often depend on the quality of the raw water, in particular on the concentration and type of organic products present.

FURTHER READING

D. C. D. Donovan, Treatment with ozone, *Journal of American Water works Association*, **57** (9), 1167 (1965)

J. R. Guillero, La micellisation–démicellisation—Nouveau procédé de traitement des eaux potables', *Travaux*, **52** (414), 444 (1969)

R. C. Sommerville et G. Rempel Ozone for supplementary water treatment, *Journal of American Water Works Association*, **64** (6), 377 (1972)

J. Mallevialle, Transformation des acides humiques par l'ozone, *Les procédés d'oxydation appliqués au traitement de l'eau potable*, Congrès de Karlsruhe, p. 307 (1978)

Y. Richard and P. Blue, Ozone pretreatment of drinking water, *Symposium de l'IOA*, Los Angelès, California (May 1978)

Y. Richard, Perspective générales d'utilisation de l'ozone, *L'eau et l'Industrie*, **31,** 35 (January 1979)

CHAPTER 20

Micellization and related effects of ozone on colloidal systems

M. R. JEKEL

DVWG–Forschungsstelle am Engler–Bunte Institut

With the application of ozone in water treatment at different locations more or less pronounced effects on turbidity, suspended solids content and particle size distributions have been observed. Terms such as micellization or micro-flocculation are often used to describe the ability of ozone to influence these parameters. Generally, a distinction may be made between direct effects of ozone on waters containing no or significant amounts of suspended matter and the indirect influence on the performance of unit processes for solids removal.

Despite the lack of knowledge in understanding all effects of ozone, it can be concluded that the oxidative action of ozone on dissolved organic substances, on organic substances adsorbed on inorganic particles and on organic solids is the responsible process.

In waters free of suspended solids, iron and manganese, but containing dissolved or colloidal organic materials such as humic substances, ozonization can lead to an increase in turbidity and suspended solids, which is not due to calcium carbonate. It is believed that the precipitation of the insoluble calcium salts of the oxidized acids is responsible for that effect. Due to the changes in the composition of the organic compounds during the year, the turbidity increase may be largely dependent on the season. It is reported that during algal activity in raw water reservoirs, ozone has had the most pronounced effects. The term micellization is often used to characterize these observations.

Surface waters containing inorganic or organic suspended solids are treated in some plants by small amounts of ozone prior to flocculation and filtration. The application of ozone alone in these preozonization units can induce an agglomeration of particles, as observed by particle size analysis. The total concentration of particles decreased up to about one-half of the original value, whereby the fraction with higher particle diameters increased. Accordingly, the turbidity of the water is often slightly reduced by preozonization. Presently, it is assumed that the protective and stabilizing action of humic acids on suspended particles is reduced by ozone treatment.

In combination with the conventional processes for removal of solids such as filtration, flocculation and sedimentation with or without filtration and flocculation–filtration, small amounts of ozone can improve the quality of supernatant or filtered water with respect to turbidity, suspended matter and particle concentrations.

These observations were made in treatment plants, where ozone was dosed into the raw water and into the clarified supernatant water. In the last case the greatest benefit of ozone was gained in a combination with flocculation–filtration, where small amounts of ferric or aluminium salts are added prior to filtration. As reported, ozonization showed the most significant improvements in flocculation with low flocculant doses. Thus ozone can be considered as a flocculant without forming additional sludge, as is the case with hydroxide flocs.

With respect to the removal of dissolved organic substances by flocculation, ozone can improve the removal efficiency, especially at low coagulant doses. The more polar oxidation products after ozonization are probably precipitated more easily by the trivalent cations or they may adsorb better on the highly polar surface of freshly precipitated hydroxide flocs. Also, the complexing ability of organics may be destroyed, as observed e.g. with ferric and manganese ions.

All the described effects of ozone may occur simultaneously in water treatment and cannot be differentiated easily. Also, it has to be checked in practice what doses of ozone are optimal with respect to the benefits and the costs.

FURTHER READING

D. Maier, Mikroflockung durch Ozon, in *Oxidationsverfahren in der Trinkwasseraufbereitung*, W. Kühn and H. Sontheimer (eds), ZfGW-Verlag, Frankfurt/Main (1979), pp. 417–41

H. Sontheimer, E. Heilker, M. R. Jekel, H. Nolte and F. H. Vollmer, The Mülheim process, *J.A.W.W.A.*, **70,** 393 (1978)

E. Wurster and G. Werner, Die Leipheimer Versuche zur Aufbereitung von Donauwasser, *gwf-Wasser/Abwasser*, **112** (81), 193 (1971)

CHAPTER 21

Interactions of ozone with chlorine and chlorine dioxide

J. VALENTA
Wasserversorgung Zürich

In present water treatment practice the combined use of two, sometimes even of three, different oxidizing agents (mainly: chlorine, ozone, chlorine dioxide) is often found. Consequently this may result in a possible reciprocal effect.

In contrast to chlorine, chlorine dioxide reacts very quickly in the water with ozone under formation of chlorate ion. These striking differences are illustrated by the UV-absorption curves shown in Figure 1. The maximum absorbance at 301 nm is proportional to the concentration of the formed chlorate. During these

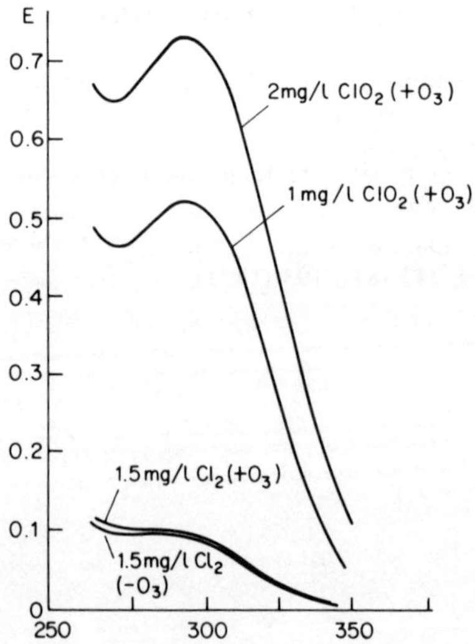

Figure 1. Chlorate formation after ozonization of the preoxidized lake water with chlorine, compared with chlorine dioxide

operational tests with normal ozone dosage of 1.5 mg/l, unusually large amounts of chlorine and chlorine dioxide were, by way of trial, added to the lake water (up to 2 mg/l of raw water).

The resulting chlorate concentration lies, in the case of dosing of gaseous chlorine, close to the detection limit of the spectrophotometric ClO_3^- method of determination. The slight chlorate formation under these circumstances confirms the second, practically overlying, curve. It corresponds to the same dosage of gaseous chlorine, but with shut-off ozone unit.

In contrast to this, the amount of chlorate created out of the chlorine dioxide is much larger, as can be gathered from the other two curves in Figure 1. With a corresponding ozone dosage the oxidation takes a stoichiometric course.

The direct UV-spectrophotometry demonstrates yet more clearly the course of the reaction between ozone and chlorine dioxide (Figure 2). Here the UV-absorption curves between 400 and 240 nm have been recorded, i.e.:

(a) ozone on its own (about 1.5 mg in 1 litre of tap water);
(b) chlorine dioxide on its own (6.6 mg in 1 litre of tap water);
(c) a mixture containing both solutions in a 1:1 proportion just after mixing—the chlorate formed in the mixture shows no distinct light absorption within this range;
(d) this is proved by the recorded curve in Figure 2, of the standard solution of 10 mg of ClO_3^- in 1 litre of tap water.

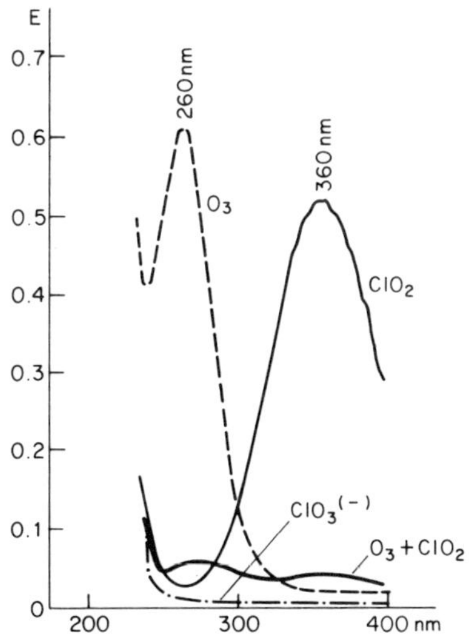

Figure 2. Changes of the UV-absorption curves as a result of the reaction between the ozone and chlorine dioxide (Device: Beckman ACTA III—Quartz cuvette dia. 50 mm, versus bidistilled water)

INFERENCES TO BE DRAWN FOR THE WATER TREATMENT PRACTICE

In aqueous solutions the oxidation of the chlorine (hypochlorites) by ozone is very slow. With normal pH values of the treated water (pH = 7) and with lower chlorine dosage (below 1 mg/l) the end-products so formed, i.e. chlorate and perchlorate, do not reach concentrations with serious hygienic or technical implications in the treatment processes. In the same conditions, the chlorine dioxide is almost immediately and completely oxidized and this reaction must be considered in view of the oxidization treatment.

FURTHER READING

T. Cheng, Spectrophotometric determination of microquantities of chlorate, chlorite, hypochlorite and chloride in perchlorate, *Anal. Chem.*, **39,** 804 (1967)

W. Koelle, Problem der gemeinsamen Anwendung verschiedener Oxidationsmittel bei der Wasseraufbereitung. *Vom Wasser*, **35,** 367 (1968)

W. J. Masschelein, *Chlorine Dioxide* (*Chemistry and Environmental Impact of Oxychlorine Compounds*). Ann Arbor Science Publication (1979)

C. J. Schack and K. O. Christe, *Inorg. Chem.*, **13,** 2378 (1974)

H. J. Schumacher, *Z. Physik. Chem.*, **13,** 353 (1957)

J. Valenta, Gemeinsamer Einsatz von Chlor und Ozon in der Wasseraufbereitung, *GWA*, **9,** 490 (1975)

CHAPTER 22

The direct action of dispersed ozonized gas bubbles

W. J. MASSCHELEIN

Compagnie Intercommunale Bruxelloise des Eaux

Fundamentally, in an ozonization process the rate-determining step is located in the reaction between dissolved molecules: ozone and substrate. Generally, oxidations with ozone are less fast in static reactors than in fully mixed contractors.

In most of the practical cases, the dissolution of ozone as a process involves gas bubbles dispersed into the liquid. The ozone diffusion in the liquid is a rather slow process [D(20°C) = 1.7410^{-5} cm/s] as compared to fast oxidations which is the case for the oxidation of iodides; for example, at least two possible ways of action can be distinguished in practice:

(a) reaction of dissolved ozone with solutes following a dimolecular scheme;
(b) fast reactions at the bubble surfaces or at the ozone-transfer layer in which the hydrophobic substances can concentrate.

The second mechanism mentioned intervenes in the preozonization phase in which no residual ozone concentration is maintained. This way of action is likely to determine the following effects:

(a) decomplexing of iron and manganese (and also other metals?) through oxidation of humic compounds;
(b) partial destruction or change in surface structure of protected colloids consecutive to the oxidation of detergents and proteins;
(c) demicellization through disrupture of the colloidal state, effect observed particularly for coloured waters;
(d) action on algae and phytoplankton;
(e) disinfection.

However, the demonstration and rational formulation of this principle remains versatile and further experimental research remains necessary. As far as the bactericidal action is concerned, the following data (see Figure 1) are obtained in a pilot plant experiment with *Actinomycetes*. Contacting systems, porous diffusors, ejector and mixing turbines are compared. Basic evidence is that the necessary ozone concentration in the water increases when the energy dissipated in the mixing is lowered. The process obeys the concept of a completely mixed reactor with a probable contact at the concentrated ozonized air bubbles.

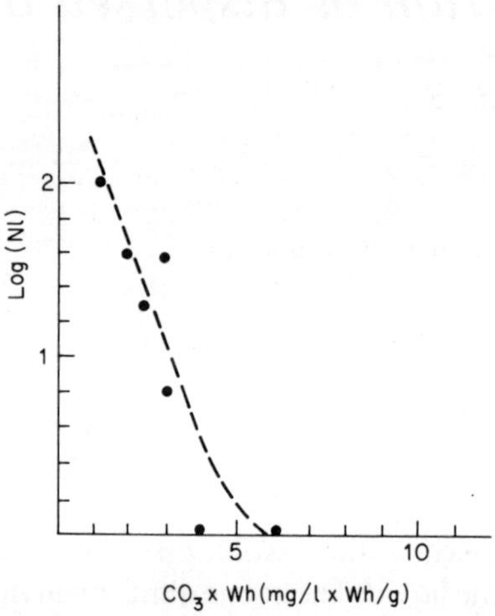

Figure 1. Influence of the mixing of ozonized air into water on the bacterial decay (Actinomycetes)

Consequently the necessary ozone doses for bactericidal action can be lowered by appropriate gas–liquid mixing energy. This model could also hold for the reactions of hydrophobic suspended matter with the concentrated ozone gas bubbles.

Further experiments have been undertaken with phenolic model compounds, e.g. picric acid, to establish the possible ozone-bubble effects on dissolved substances to be oxidized. Picric acid has been chosen as a model substance to be ozonized as being representative of a slow-reacting phenol. It is specifically determined by polarography (see Figure 2).

In the experimental set-up the porosity effect of the diffuser, the bubble size and number have been investigated. The column was operated in static liquid conditions in distilled water so that only the gas expansion and bubbling contribute to mixing. The bubble size and rising velocity are estimated by photography and confirmed by measuring the volumic expansion of the liquid when ozonized air is injected in the static water layer.

In the 'static' diffusion systems the most active bubble size is that of diameter 4.5 mm with a velocity gradient at the bubble-formation point of 7.400 s^{-1}.

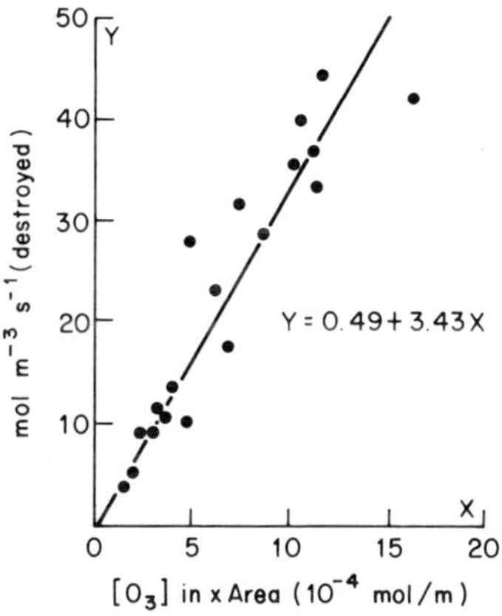

Figure 2. Global kinetics of ozonization of picric acid and ozone transfer in distilled water

The ozonization rate is proportional to the product of the ozone concentration in the gas phase and the surface of this phase. This equally holds for the mass transfer of the ozone from the gas to the liquid phase which is consequently believed determinant in the rate of the process. As the maximum transfer occurs during the bubble formation and maximum velocity gradient in practical systems, the mixing of the liquid which is intended to ensure repeated gas–liquid contact, favours the oxidation rate.

FURTHER READING

W. J. Masschelein, Contacting of ozone in water including off-gas treatment. In *Opus Ozone*, R. G. Rice and A. Netzer (eds), Ann Arbor (in preparation)

W. J. Masschelein, *Belgian Experiences in Ozonization of Water*. IOA Congress, Berlin (1981) p. 433 in Colloquium Proceedings, O. H. Hess, Berlin (1981)

W. J. Masschelein and R. Goossens (in preparation)

CHAPTER 23

Perspective of combination of ozone and ultrasound

E. DAHI

Technical University of Denmark

In the field of waterworks, the advantage of a combination of two agents to obtain more favourable action than if these were used separately, is obvious. By itself, water is a poor cleaner, soap is not a cleaner at all, but the combination of the two provides a convenient, efficient, rapid and economical cleansing. Yet disinfection of water is based on the separate use of few single agents. These 'poor cleaners' have been investigated and utilized for a period of 100 years, but the beneficial 'soaps' remain to be researched and developed. This chapter deals with one of the potential combinations: sonozone.

SYNERGISTIC EFFECTS

In gas/liquid semiflow laboratory experiments, ultrasonic treatment, 10 W/l, 20 kHz, was able to increase the rate constants of ozone oxidation of rhodamine-B by 37–55% (Figure 1). Similar synergisms were reported for reduction of chemical oxygen demand and total organic carbon. The liquid/liquid technique and the non-simultaneous combination of ultrasound and gaseous ozone are not capable of reproducing these effects.

The employed ultrasonic treatment in itself does not cause significant inactivation of virus and bacteria. On the contrary, increase of the microbial count has been reported. Nevertheless, pre- as well as simultaneous sonication are able to bring about a synergistic action with ozone in the gas/liquid and, to less extent, in liquid/liquid techniques (see Table 1). Seventy per cent of the ozone dosage required for 3 decades inactivation, and 80% of the ozone dosage required for 4 decades inactivation by gaseous ozone alone, could be saved by the simultaneous application of ultrasound in steady-state laboratory experiments (see Figure 2).

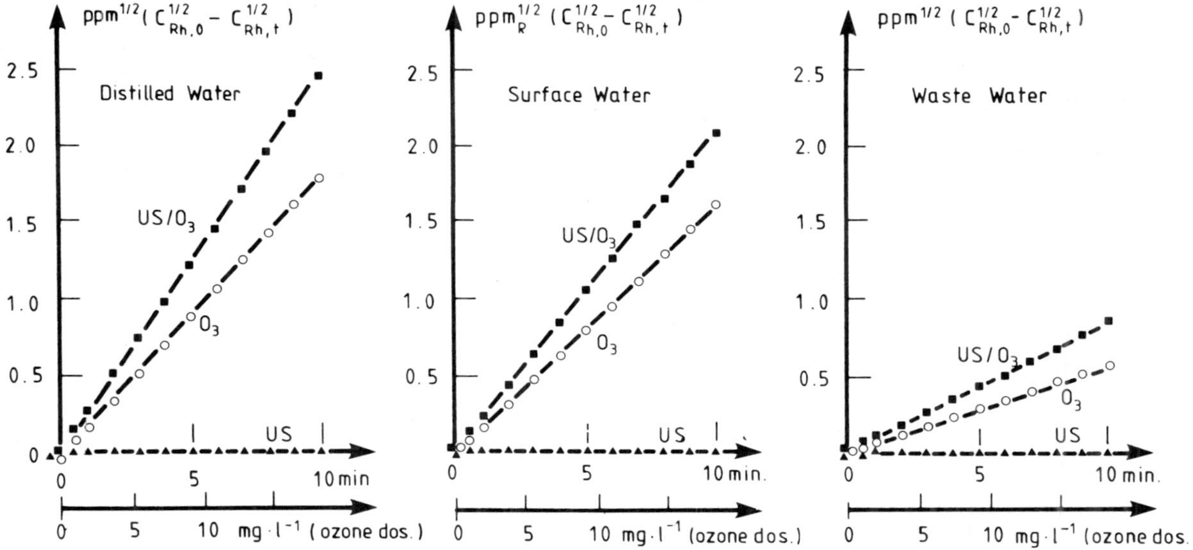

Figure 1. Decolorization of rhodamine-B *vs.* time with ultrasonic treatment, with gaseous ozone dosage, and with simultaneous combination of the two in semiflow experiments

Table 1. Division and characteristics of the function of inactivation by ozone with and without ultrasound

Curve segment	Characteristic	Suggested nomenclature	Explanation
Stage 1	Almost horizontal	Ozone demand stage	Prolonged by addition of ozone consuming pollutants and by presonication of polluted water. Reduced by simultaneous sonication in gas/liquid techniques.
Stage 2	Steep slope, deflection tangent	Proper inactivation stage	Apparently not affected by ultrasound. Decreased slope in flow technique and by very slow ozone dosages in semiflow technique.
Stage 3	Slow slope	Aggregate inactivation stage	Eliminated or reduced by sonication and by filtration prior to ozonization.

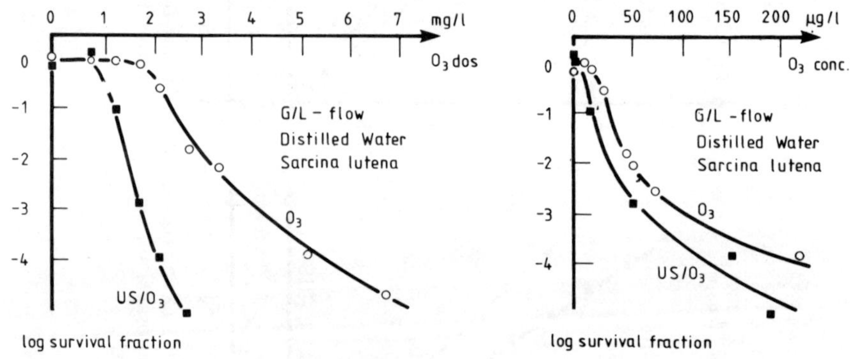

Figure 2. Steady-state inactivation of bacteria by gaseous ozone with and without ultrasonic treatment

Thus, in combination the effect of ultrasound and ozone is synergistic chemically as well as microbiologically. Under certain circumstances, however, the combination effects may be additive or antagonistic.

INTERACTIONS AND MECHANISMS

The pattern of interactions between ozone and ultrasound is very complex, and not yet completely understood. In the ozonization processes, gaseous ozone is transferred to the water and the dissolved ozone reacts with the contents of dissolved chemicals, suspended solids and micro-organisms. Dissolved ozone also produces free radicals, some of which are able to initiate chain reactions with ozone and to oxidize dissolved chemicals and probably to inactivate microorganisms (see Figure 3).

Propagation of intensive ultrasonic waves in the water brings out the interference phenomenon called cavitation. Cavitational ultrasound is able to act on water and, via a free radical mechanism, to produce peroxide at mg/m^3 levels. It is also capable of breaking chemical bonds and of transforming small amounts of a wide variety of dissolved chemicals.

Cavitational ultrasonic irradiation of dissolved ozone increases the ozone autodecomposition significantly, presumably by back-feeding its chain transfer

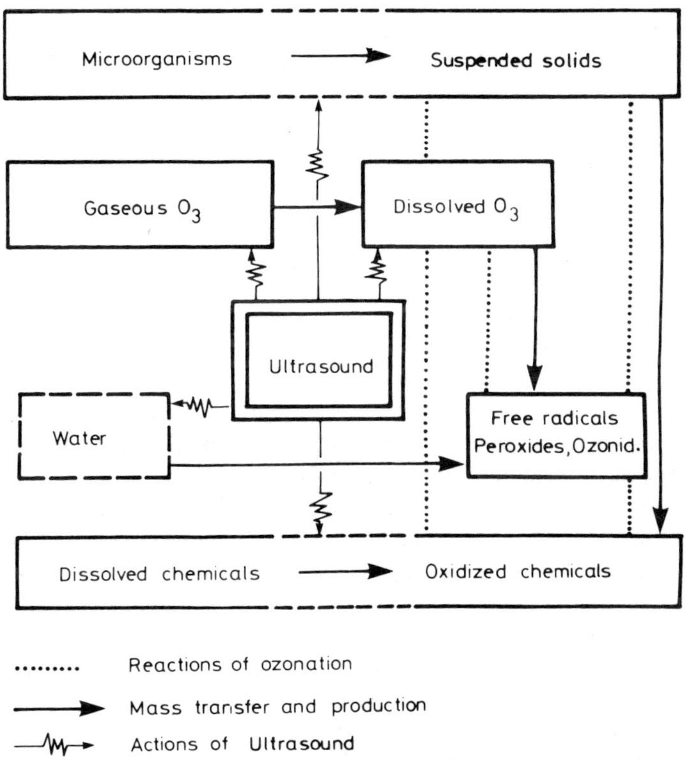

Figure 3. Simplified schematic diagram of the processes and interactions of sonozone

to free radicals. Furthermore, the direct molecular ozone reactions may be accelerated due to the local heat and pressure effects of cavitation. The vibration energy is, in itself, not great enough to cause chemical changes.

Ultrasound acts on ozone bubbles and increases the mass transfer between gas and the liquid. This effect has been explained as a matter of disruption of the ozone bubbles and of expansion of the gas/liquid contact surface. Indeed, small bubbles often increase in size to resonate. At the resonance, the microscopic turbulence of the cavitational and/or vibratory energy increase the gas/liquid mass transfer, apparently even though the macroscopic turbulence of some mixing device is already high.

The action of ultrasonic irradiation is reported to have several effects on microorganisms in water. Both growth and inactivation may take place depending upon, among other things, the ultrasonic intensity and the duration of the treatment. At the intensities and residence times, which may be realistic for water treatment, the disaggregation of microbial flocs is of major importance.

Ozone is very efficacious and rapid to perform the first few decades of inactivation, but is inert and slow to inactivate the residual microbial floc, which is encapsulated. Surviving bacteria in flocs may be protected by refractive organic matter, which is made biodegradable by ozone itself. Thus, at the time where ozone has disappeared, the encapsulated infectious material will be vital and cause aftergrowth. For this reason sonozone, by disaggregation of flocs while ozone is still present, may give a more efficient inactivation and reduce the aftergrowth problems of water works.

METHODS AND TECHNOLOGY

Our knowledge on the methods and technology of sonozone is extremely limited. No results have been reported on sonozonization of authentic untreated or partly treated drinking water; neither in the laboratory, nor in pilot plant equipment. Indeed, the techniques employed have been chosen more or less blindly. The sonozonization technology of today is to be compared with the ozonization technology of the beginning of this century.

Ultrasounds are usually produced by piezoelectric transducers connected with a generator. The anisotropic crystals of the transducers, for example barium titanate, are exposed to a reversing electric field. This makes the crystals expand and contract periodically, and generate the acoustic waves in the adjacent medium.

It is evident that ultrasounds and ozone can be combined in many different ways. Presumably the simultaneous treatment in directly injected ozonization shafts is the most promising method (Figure 4). Even though most experiences have been gained at 20 kHz, ultrasound is believed to act synergistic at a very wide range of frequencies. The intensity is supposed to be of greater importance. Nett

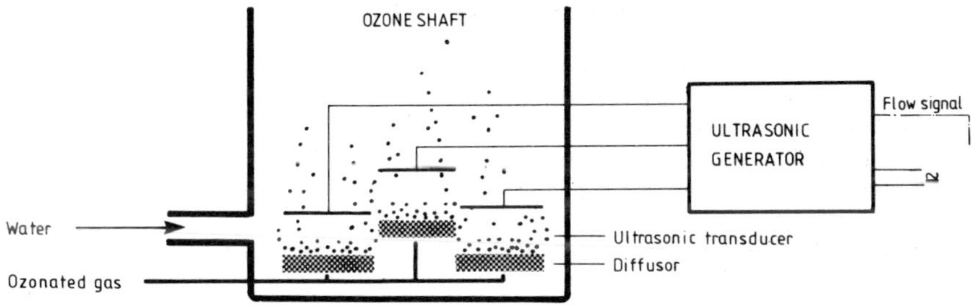

Figure 4. Suggested sonozone equipment

intensities of the order of 1 W/cm^2 have been employed successfully in the laboratory experiments. Apparently, maximum effects would be obtained if the cavitational ultrasonic waves are focused closely over the site of initial contact between ozone bubbles and the flowing water. In such a system the water flow can be increased beyond what has until now been attempted.

ECONOMICS AND FURTHER DEVELOPMENTS

At waterworks dose levels, the effects of ultrasound have been evaluated as follows: if the ozone dose Do_3 is required to obtain a given effect by ozonization, and a smaller ozone dose do_3 is required by an equal sonozonization at which the ultrasound dose d_{US} is employed, then:

$$UE_{sonozone} = 2.6\,[(1 - SF).\,Do_3]^{05} + 0.14\,d_{US}^{09}$$

where: UE is the uniform expense;
D,d is the dose in mg O_3/l or Wh ultrasound/m^3;
SF is the ozone saving factor $= 1 - do_3/Do_3$.

A sonozonization with assumed 50 Wh/m^3 ultrasonic dosage and 0.7 saving factor, was estimated to be slightly more expensive than ozonization alone. Thus, the effective synergism observed seems, at the moment and from an economical point of view, not to be attractive for waterworks.

Of course, this balance sheet would easily be turned around if the laboratory results cannot be reproduced at a technical scale. On the other hand, technical development by optimization of the employed ultrasonic intensity and frequency, or by production of less expensive and more durable transducers, would invert the balance.

However, toxicological, microbiological, chemical or technical aspects, about which we still know very little, may be of greater importance than the economic parameters for the choice of disinfection methods at waterworks.

FURTHER READING

G. R. Burlesson, T. M. Morray and M. Pollard, *Appl. Microbiol.*, **29,** 340–4 (1975)
E. Dahi, *Wat. Res.*, **10,** 677–84 (1976)
E. Dahi, The 3rd Congress of the Int. Oz. Inst., IOA, Paris (1977)
E. Dahi, Int. Ozone Ass., European Commit. Workshop, IOA, Nice (1978)
E. Dahi and E. Lund, Presented at the 4th World Ozone Congress of the Int. Oz. Inst., Houston (1979)
R. A. Sierka, The 3rd Congress of the Int. Oz. Inst., IOA Paris (1977)

CHAPTER 24

Scope of joined action of ozone and ultraviolet light

W. J. MASSCHELEIN

Compagnie Intercommunale Bruxelloise des Eaux

On the contact of ozone with dissolved organic substances in water at least two different reactional pathways occur: the first, an attack of the unsaturated sites under addition of ozone forming on hydrolysis an added hydroperoxide group; the second involves the formation of a fast-reacting $OH^{\cdot}$ radical intervening through a direct reaction (see Chapter 11).

Part of the action of UV radiation, sometimes used on itself for the disinfection of water, is attributed to the inside formation of free radicals like $O_2H^{\cdot}$, $HO^{\cdot}$ and even $O^{(3p)}$. These species appear directly related to the reactional mechanisms of ozone, thus justifying the attempts to associate ozone and UV radiation.

It must be emphasized in this context that, if UV irradiation of oxygen (or oxygen in air) can built-up ozone the reactions necessary therefore are not actuated in aqueous medium. As a whole, the reactions implicated in the joint use of ozone and UV radiation are part of the photochemistry of oxygen reviewed earlier.[1] Part of the over-all action may also result from the activation of the compounds to be oxidized by the action of the photons introduced into the medium. Whence, the technique is, at first sight, directly suited for the oxidation of refractory compounds which are however capable of forming free radicals. Among these: cyanides (inclusive complex cyanide ions), alcohols and also chlorinated compounds, particularly THM's.[2,3] Brominated derivatives react faster comparatively to similar chlorinated compounds.

Also in the disinfection of heavily polluted waters, including waste-water, the joint action of ozone and UV radiation enables an increase in disinfection efficiency by a factor of ten when compared to ozone alone.

On a global basis the mechanism can be summarized as follows (S being the substrate and R the reaction product):

$$O_3 + h\nu \rightarrow O_2^* + O^{\cdot} \qquad S^{\cdot} + H_2O \rightarrow 2\,OH^{\cdot}$$

$$O^{\cdot} + H_2O \rightarrow 2\,OH^{\cdot} \qquad S^{\cdot} + O_3 \rightarrow O_2^* + O^{\cdot}$$

$$OH^{\cdot} + S \rightarrow R \qquad \dot{O}_3 + OH \rightarrow HO_2 + O_2$$

$$S + h\nu \rightarrow S^{\cdot}$$

As a general rule it can be considered that the global reaction is of first order in substrate concentration and its velocity depends both on the intensity and the doses applied. Furthermore, the reactions are often slower at higher pH values. At present, the simultaneous use of UV/O_3 is still in its early stages and the technology of the contact chambers to treat important water flows is still to be developed.

In the experimentation described this far, static diffusion of the ozonated gas introduced into the contacting reactors is used. The reactors are equipped with a fast mixer and immersed UV lamps. The largest reactors mentioned at present have an active volume of $25\,m^3$. However most of the experiences are still developed on the laboratory scale and use reactors of a few litres capacity. The literature mentions UV power capacities up to $1.5\,kW/m^3$ of reactor volume! Fortunately more recent data report applications of $0.2\,kW/m^3$. The dissipated energy ranges at present from 0.1 to $0.4\,kW/m^3$ and depends on the reactivity of the substrates. The UV energy consumption amounts 15 to 40 kJ/mole.

The necessary reaction time to alter and destroy refractory compounds as the THM is of at least 30 min.

The actual and preliminary conclusion is that with the available technology the UV/O_3 method implicates a cost of about double that of conventional ozonation.

REFERENCES

1. W. J. Masschelein, *T.S.M.-L'eau*, **77**, 177 (1977)
2. H. W. Prengle and Ch. E. Mauk, *Am. Inst. Chem. Eng. Symp.*, Ser. 1978, **74**, 228 (1978)
3. H. W. Prengle, C. G. Hewes and W. E. Mauk, *Comm. 2nd. Intern. Ozone Symp.*, Montreal (1975)
4. G. R. Peyton, J. L. Buleson, F. Huang, S. Lin and W. H. Glaze, *Comm. 4th World Ozone Congress*, Houston (1979)

CHAPTER 25

Identified reaction products in ozonization

J. MALLEVIALLE
SLEE

This list of substances has been obtained from data given in the literature and is therefore not claimed to be exhaustive. It relates to substances obtained in an aqueous medium without reference to the specific reaction conditions (pH, presence of a catalyst and so on).

The first column of the table sets out first the compounds (C) for which certain ozonization by-products are known (given in column 2) and secondly those compounds which are themselves by-products of the ozonization of other substances shown in column 3.

Compound (*C*)	Ozonization by-products of (*C*)	(*C*) Precursors by ozonization
ACETALDOXIME		Diethylamine
ACETIC ACID	Glyoxylic acid Oxalic acid Diacetyl	Acetone Cresol 4-Chloro-*o*-cresol Ethylic alcohol Mesitylene Methylglyoxal Methylethylketone Phenol Xylenol
ACETIC ALDEHYDE		Ethylic alcohol Linolenic acid
ACETONE	Acetic acid Formic acid Formaldehyde Oxalic acid	Isopropanol
ACETYLENIC ACIDS	Carboxylic acid	
ADIPIC ACID		Cyclohexanol Cyclohexene
ALBUMIN from blood (ox)	Ammonia Nitrates	
ALCOHOLS		Ethers
ALCOHOL(III) tertiary		Alkylmercurichalides

ALDEHYDES	Carboxylic acids	Alkenes Aliphatic alcohols Humic acids insaturated Olefins
ALDRIN	Dieldrin	
ALIPHATIC ALCOHOLS	Aldehydes Carboxylic acids Ketones	
ALKENES	Aldehydes Hydroxyhydroperoxides Carboxylic acids Ketones	
ALKYLBENZENE SULFONIC ACID	Formaldehyde Formic acid Alkylglyoxal	
ALKYLGLYOXAL		Alkylbenzene sulfonic acids
ALKYLMERCURICHALIDES	Carboxylic acids Ketones III-Alcohol	
AMINES I and II (primary and secondary)	Nitrocompounds	
AMINO-2-BENZIMADAZOL	Hydroxy-2-benzimidazol Benzimidazolylamino-2-benzimidazol S-Triazin	
4-AMINOBENZOIC ACID	Ammonia Formic acid	

Compound (*C*)	Ozonization by-products of (*C*)	(*C*) Precursors by ozonization
	Nitrocompounds Nitrates Oxalic acid	
AMMONIA	Nitrates	Albumin from blood (ox) 4-Aminobenzoic acid Histidine Leucine
ANTHRACENE	Anthraquinone Phtalic acid	
ANTHRAQUINONE		Anthracene
BENZALDEHYDE	Benzoic acid	Styrene
BENZANTHRACENE	Hydrogen peroxide Phtalic acid Quinone	
BENZENE	Glyoxal Glyoxylic acid	
BENZIMIDAZOLYLAMINO-2-BENZIMIDAZOL		Amino-2-Benzimidazol
BENZOFURAN	Pyrocatechol Salicylic acid Salicylic Aldehyde	
BENZOIC ACID		Benzaldehyde Styrene

BROMIDE	Hypobromous acid	
BIS(CHLORO-6-OXO-2-BENZOXAZOLYL METHYL) ETHER		Phosalone
t-BUTYLAMINE	Nitroalkanes	
CAFFEINE	Dimethylparabanic acid *N*-Methyloxalamide *N*-*N*′-Dimethyloxalamide	
CARBOXYLIC ACIDS	Carboxylic acids	Acetylenic acids Aldehydes Aliphatic alcohols Aliphatic Aldehydes Alkylmercurichalides Carboxylic acids Ketones
CARBOXYLIC POLYACIDS		Humic acids
CHLORIDES	Chlorine Hypochlorous acid	4-Chloro-o-cresol Chlorophenol Dichlorophenol 2,3,4,5-Tetrachlorophenol Trichlorophenol-2,4,6 2,4,5-T acid
CHLORINE		Chlorides
4-CHLORO-*O*-CRESOL	Chlorides Acetic acid Formic acid Methylglyoxal	

Compound (*C*)	Ozonization by-products of (*C*)	(*C*) Precursors by ozonization
	Oxalic acid Pyruvic acid	
o- *m*-CHLOROPHENOL *p*-	Chlorides Glyoxylic acid Glyoxal Hydrogen peroxide *p*-, *o*- Quinone	
CRESOL	Acetic acid Glycolic acid Glyoxylic acid Maleic acid Oxalic acid Propionic acid	
CROTONALDEHYDE	Acetaldehyde Glyoxal Formic acid	
CUMENE	Cumene peroxide	
CUMENE PEROXIDE		Cumene
CYANATES		Cyanides Thiocyanates
CYANIDES	Cyanates	Thiocyanates
CYCLOHEXENE	Adipic acid Glutaric acid	

CYCLOHEXANOL	Malonic acid Succinic acid	
CYSTEIC ACID		Cystine
DDE	DDT by-products	DDT
DDT	Hydroquinone (para) Maleic acid Muconic acid Phosgene DDE	
DECYLBENZENESULFONIC ACID	Formic acid Formaldehyde Nonylglyoxal	
cis-1,2-DIBENZOYLETHYLENE		2,5,Diphenylfuran
DICHLOROMALEIC ACID		2,4,5-T acid
DICHLOROPHENOL	Chlorides and phenol by-products	
DIELDRIN		Aldrin
DIETHYLAMINE	Acetaldoxime	
DIETHYLCHLOROPHOSPHATE	Tetraethylpyrophosphate	Diethylchlorothiophosphate
DIETHYLCHLOROTHIOPHOSPHATE	Diethylchlorophosphate Tetraethylpyrophosphate	
DIETHYLDITHIOPHOSPHORIC ACID	Phosphoric acid Disulphide compound Trisulphide compound	

Compound (*C*)	Ozonization by-products of (*C*)	(*C*) Precursors by ozonization
DIHYDROXYFUMARIC ACID	Dihydroxytartaric acid Mesoxalic acid Mesoxalic semi-aldehyde acid Oxalic acid	
DIHYDROXYPHENYLALANINE		Tyrosine
DIHYDROXYTARTARIC ACID		Dihydroxyfumaric acid
N,*N*′-DIMETHYLOXALAMIDE		Caffeine
DIMETHYLPARABANIC ACID		Caffeine
2,4-DINITROPHENOL		Paraoxon Parathion
DIPHENIC ACID		Phenanthrene
2,2-DIPHENYLALDEHYDE		Phenanthrene
2,5-DIPHENYLFURAN	*cis*-1,2 Dibenzoylethylene	
DIPHENYLHYDRAZINE	Anilinophenol	
EDTA—Chelated	Metallic ions	
EPOXYDES		Unsaturated olefins
ESTERS		Ethers
ETHERS	Alcohols Esters	

ETHYLIC ALCOHOL	Acetic acid Acetaldehyde Formic acid Hydrogen peroxide	
FORMALDEHYDE		Acetone Alkylbenzenesulfonic acid Decylbenzene sulfonic acid Nonylbenzenesulfonic acid Phenol Styrene
FORMIC ACID		Acetone Alkylbenzenesulfonic acids 4-Aminobenzoic acid 4-Chloro-*o*-cresol Crotonaldehyde Decylbenzene sulfonic acid Ethylic alcohol Humic acids Maleic acid Mesitylene N-Methyldithiocarbamate dihydrate Na Methylethylketone Naphtalene,2,7,-disulfonic acid Nonylbenzenesulfonic acid Oxalacetic acid Phenol
N-FORMYLKINURENINE		Tryptophan
FULVIC ACIDS	Benzaldehyde Acetic acid 2-Methylphenyl-ester	

Compound (*C*)	Ozonization by-products of (*C*)	(*C*) Precursors by ozonization
	Isobenzofurandione Benzene 1,1-(1,2-Ethanediyl) bis ester Methylpalmitate Nicotinonitrile Dibutylphtalate Dimethylheptadecanoate Dicyclohexylphtalate Bis-2-ethyl hexylphtalate Methylcyclohexane Isobenzofurandione 4-methylbenzoic acid Dioctylphtalate	
FUMARIC ACID		Muconic acid
GLUTARIC ACID		Cyclohexane
GLYCINE		Isopropoxyphenyl methyl-carbamate (Baygon)
GLYCOLIC ACID		Cresol Humic acids Maleic acid 2,4,5,-T-Acid Xylenol
GLYOXAL	Glyoxylic acid Oxalic acid	Benzene 2-Butenal Naphthtalene Chlorophenol

		Crotonaldehyde 2,7-Naphthalene disulfonic acid Phenol
GLYOXYLIC ACID	Oxalic acid	Acetic acid Benzene Chlorophenol Cresol Glyoxal Oxalacetic acid Phenol 2,4,5,T-acid Xylenol
HEPTACHLOR	Heptachlorepoxide	
HEPTACHLOREPOXIDE		Heptachlor
HEXANALDEHYDE		Linoleic acid
HISTIDINE	Ammonia	
HUMIC ACIDS	Aldehydes Formic acid Glyoxylic acid Carboxylic polyacids Ketones Oxalic acid	
HYDROGEN PEROXIDE		Benzanthracene Orthochlorophenol Ethylic alcohol Malonic acid Naphthalene

Compound (*C*)	Ozonization by-products of (*C*)	(*C*) Precursors by ozonization
		Phenanthrene Phenol
HYDROQUINONE		DDT Phenol Cresol
HYDROXY-2-BENZIMIDAZOLE		Amino-2-benzimidazol
HYDROXYHYDROPEROXYDES		Alkenes
HYDROXYMETHYL-3-CHLORO-6 OXO-2-BENZOXAZOLONE		Phosalone
HYPOBROMOUS ACID		Bromide
IODIDES	Iodine	
IODINE		Iodides
ISOPROPA	Acetone	
O-ISOPROPOXYPHENYLMETHYL CARBAMATE (Baygon)	Glycine	
KETONES	Carboxylic acids	Alkenes Aliphatic alcohols Alkylmercurichalides Humic acids

LINOLENIC	Propionic acid Propionaldehyde Acetaldehyde	
LEUCINE	Nitrates Ammoniac	
LINOLEIC ACID	Hexanaldehyde Propionic acid Propionaldehyde	
MALATHION	Malaoxon Phosphoric acid	
MALEIC ACID	Formic acid Glyoxylic acid Oxalic acid	Cresol DDT Muconic acid Xylenol
MALONIC ACID	Hydrogen peroxide Mesoxalic acid Tartronic acid Oxalic acid	Cyclohexanol
MALONIC ALDEHYDE		Methylinoleic acid
MALAOXON		Malathion
MESITYLENE	Acetic acid Formic acid Methylglyoxal	
MESOXALIC ACID		Dihydroxyfumaric acid Malonic acid

Compound (*C*)	Ozonization by-products of (*C*)	(*C*) Precursors by ozonization
		Naphthalene-2,7,-disulfonic acid Oxalacetic acid
MESOXALIC SEMIALDEHYDE		Dihydroxyfumaric acid Naphthalene-2,7,-disulfonic acid
METHANOL		*N*-Methyldithiocarbamate dihydrate Na (Vapam)
METHIONINE	Methionine sulfoxide	
METHIONINE SULFOXIDE		Methionine
N-METHYLDITHIOCARBAMATE DIHYDRATE Na (VAPAM)	Methanol Formic acid	
METHYLETHYLKETONE	Acetic acid Formic acid Propionic acid	
METHYLGLYOXAL	Acetic acid Pyruvic acid	4-Chloro-*o*-cresol Mesitylene
METHYLLINOLEIC ACID	Malonic aldehyde	
N-METHYLOXALAMIDE		Caffeine
MUCONIC ACID	Fumaric acid Maleic acid Oxalic acid	DDT Phenol

MUCONIC ALDEHYDE		Phenol
MESOXALIC SEMIALDEHYDE ACID		Dihydroxyfumaric acid Naphthtalene-2,7,6 disulfonic acid
NAPHTHALENE	Phtalic aldehyde Hydrogen peroxide Phtalic acid Phtaldehyde acid	
NAPHTHALENE-2,7-DISULFONIC ACID	Glyoxal Formic acid Mesoxalic acid Mesoxalic semialdehyde acid Oxalic acid	
NICOTINIC ACID		Quinoline
NITRATES		Albumin from blood (ox) 4-Aminobenzoic acid Ammonia Leucine 2-Nitro-p-cresol Urea
NITROCOMPOUNDS		Amines I and II (primary and secondary) 4-Aminobenzoic acid
2-NITRO-*p*-CRESOL	Nitrates	
NONYLBENZENESULFONIC ACID	Formaldehyde Formic acid Nonylglyoxal	

Compound (*C*)	Ozonization by-products of (*C*)	(*C*) Precursors by ozonization
NONYLGLYOXAL		Decylbenzene sulfonic acid Nonylbenzenesulfonic acid
OCTANOIC ACID		Oleic acid
Unsaturated OLEFINS	Aldehydes Epoxydes Enols (vinylics alcohol)	
OLEIC ACID	Octanoic acid	
OXALACETIC ACID	Formic acid Glyoxylic acid Mesoxalic acid Oxalic acid	
OXALIC ACID		4-Chloro-*o*-cresol Cresol Acetic acid Acetone Dihydroxyfumaric acid Glyoxal Glyoxylic acid Humic acids Maleic acid Malonic acid Muconic acid Naphthalene-2,7,-disulfonic acid Oxalacetic acid Phenanthrene Xylenol

PARAOXON	2-4-Dinitrophenol Picric acid	Parathion
PARATHION	Paraoxon 2-4-dinitrophenol Phosphoric acid Picric acid	
PHENANTHRENE	Diphenic acid 2-2 Diphenylaldehyde Hydrogen peroxide Oxalic acid	
PHENANTHRENE QUINONE		9-Phenanthrene carboxylate Na
9-PHENANTHRENE CARBOXYLATE Na	Phenanthrenequinone	
PHENOL	Aldehyde muconic acid Pyrocatechol Hydroquinone Glyoxylic acid Resorcinol Muconic acid Oxalic acid Formic acid Acetic acid Glyoxal Formaldehyde Muconaldehyde Hydrogen peroxyde	Salicylic acid
PHOSGENE		DDT
PHOSPHATE		Phosphite

Compound (*C*)	Ozonization by-products of (*C*)	(*C*) Precursors by ozonization
PHOSPHORIC ACID		Diethyldithiophosphoric acid Malathion Phosalone
PHOSALONE	Bis-(Chloro-6-oxo-2-benzoxazolyl-methyl-3) ether Hydroxymethyl-3-chloro-6-oxo-2-benzoxalone Phosphoric acid	
PHOSPHITE	Phosphate	
PHTALIC ACID		Anthracene Naphthalene
PHTALIC ALDEHYDE		Naphthalene
PHTALDEHYDE ACID		Benzanthracene Naphthalene
PICRIC ACID		Paraoxon Parathion
PROPIONALDEHYDE		Linoleic acid Linolenic acid
PROPIONIC ACID		Cresol Linoleic acid Linolenic acid Methylethylketone Xylenol

PYROCATECHOL		Phenol Benzofuran Salycylic acid
PYRUVIC ACID		4-Chloro-*o*-cresol Methylglyoxal
o-QUINONE *p*-QUINONE		Orthochlorophenol Phenol
QUINOLINIC ACID		Quinoline
QUINOLINE	Quinolinic acid Nicotinic acid	
RESORCINOL		Phenol
SALICYLIC ACID	Phenol Pyrocatechol	Benzofuran
SALICYLALDEHYDE		Benzofuran
SELENIDES	Selenoxides	
STYRENE	Benzaldehyde Benzoic acid Formaldehyde	
SUCCINIC ACID		Cyclohexanol Cyclohexene
SULPHONES		Sulphoxides Sulphides (R_2-S)

Compound (*C*)	Ozonization by-products of (*C*)	(*C*) Precursors by ozonization
SULPHOXIDES	Sulphones	Sulphides (R_2-S)
SULPHIDES (R_2S)	Sulphoxides Sulphones	
2-4-5-T ACID	Chlorides Dichloromaleic acid Glyoxylic acid Glycolic acid Oxalic acid	
TARTRONIC ACID	Mesoxalic acid	Malonic acid
2,3,4,5-TETRACHLOROPHENOL	Chlorides Phenol by products	
TETRAETHYLPYROPHOSPHATE		Diethylchlorophosphate
Substituted TETRA N-PHENYLENEDIAMINE	Würster's salt	
THIOCYANATES	Cyanides Cyanates	
s-TRIAZINE		Amino-2,-benzimidazole
2,4,6-TRICHLOROPHENOL	Chlorides Phenols by-products	
TRYPTOPHAN	N-Formylkinurenin	

TYROSINE	Dihydroxyphenylalanine	
UREA	Nitric acid Nitrates	
VINYLIC ALCOHOLS		insaturated OLEFINS
XYLENOL	Acetic acid Glyoxalic acid Glycolic acid Maleic acid Propionic acid Oxalic acid	

3. PRACTICAL ASPECTS OF OZONIZATION

CHAPTER 26

Significance of the ozone to water contacting system

M. BREDTMANN

Gebr. Hermann Köln

The use of ozone in water treatment is often based on the concept of the reaction of dissolved ozone with water-soluble substrates. Hence the ozone transfer from the gas phase into the water to be treated is an important process. Therefore the equipment available for mass transfer processes are packed columns, plate columns, gas bubbling contactors, ejectors, turbines and porous columns. The choice between these possibilities involves severe reference to each particular case as the 'best' reactor generally must be selected in function of the task to be fulfilled by the treatment as a whole. The aim of this contribution is to give a rough survey of the specification to be met by an ozone-to-water contactor when based on chemical and technological requirements.

Ozone-contacting as a whole is the sum of a series of partial processes which can be described in a simplified way by the transfer-phenomena (from the gas phase to the reacting molecules) and the reaction itself with the existing dissolved compounds.

The distribution of the ozone introduced can be represented schematically as in Figure 1, from which can be deduced the unavoidable minimal ozone loss in the vent gas resulting from the distribution of the transferred ozone. In this context, the fundamental equation for the mass transfer given by

$$m_{O_3} = \beta . F_A . \Delta C$$

illustrates the fundamental conditions to be fulfilled by the reactor. In order to realize the maximum transfer of ozone, the phase exchange coefficient, β, the interface between phases, F_A and the saturation deficit ΔC must be as high as possible. Obtaining the necessary contact surface to promote a sufficient phase transfer does not give rise to major problems with the known reactors applied in present technologies. In a single-stage process the deficit in concentration compared to the saturation value, ΔC, is sufficient as driving force for the mass transfer, seeing that the initial dissolved ozone concentration is zero. The ozone to water contact is favoured by a counter-current action in a contactor without important recirculation of the liquid.

The maximal dissolved ozone concentration can be computed according to Henry's law on the gas solubilities expressing explicitly the proportionality of the saturation concentration of a dissolved gas to the partial pressure:

$$C_{F(\max)} = C_G \cdot A$$

In other words the soluble quantity of ozone giving the concentration C_F is determined by the concentration in the gas phase C_G and the general solubility A of the theoretically pure gas (e.g. 100 % ozone). In practice, this relation means that the ozone concentration in the gas coming from the ozonators can influence the efficiency of a contacting reactor.

In Figure 1, curve number 1 illustrates schematically the proportionality expressed by Henry's law. Curve number 2 illustrates the behaviour of a water charged with fast-reacting impurities with ozone, while curve number 3 indicates

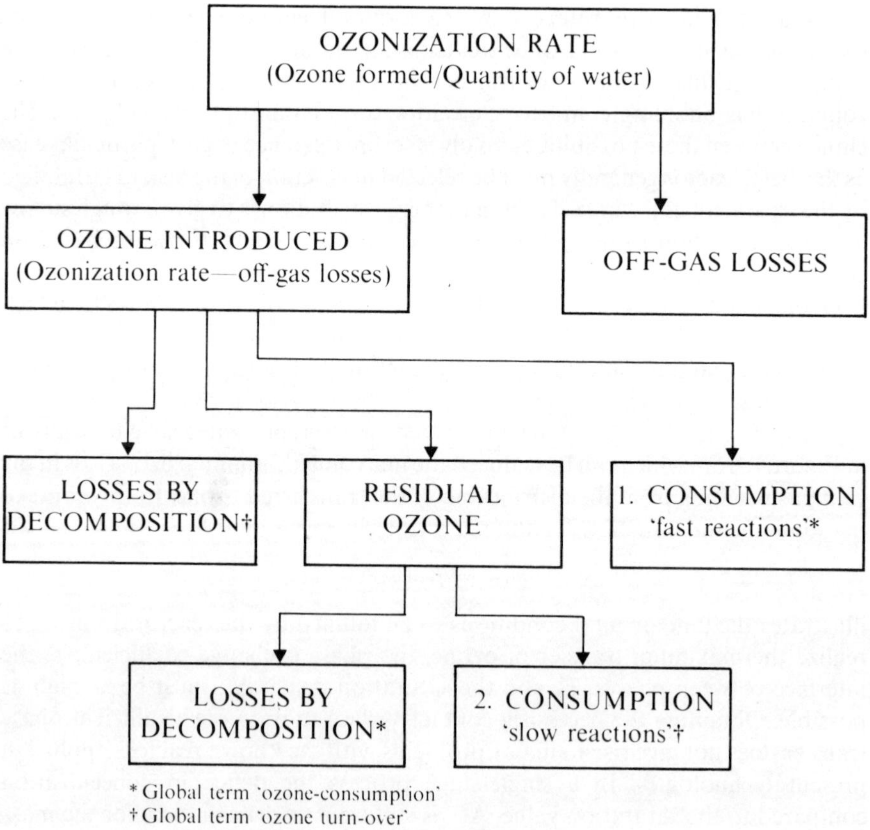

* Global term 'ozone-consumption'
† Global term 'ozone turn-over'

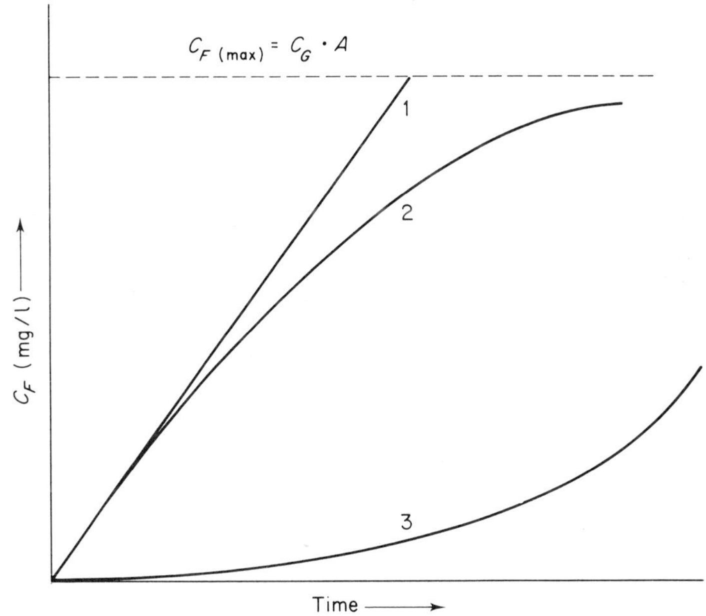

Figure 1

the presence of slow-reacting compounds. These experimental records enable us to deduce the influence of the nature of oxidable solutes on the phase exchange coefficient 'β' which is increased when fast-developing reactions occur: the dissolution is chemically assisted. In short, the ozonator releasing a given ozone concentration in the gas phase, and the reactor, must be designed as an integrated system, and optimized for each specific ozonization problem.

Another and quite different question is to limit to a minimum extent the ozone losses due to ozone decomposition. Direct measuring of these losses is practically impossible but a long-range balance can reveal precise information on the subject.

The global balance in which all the data on ozone concentration (C_F) are reported on the basis of the volume of water treated are as given in Figure 2. At lower water flow the amount of ozone required for the reactions is significantly higher, although practically, no further improvement in the water quality is obtained. To avoid or limit such ozone losses the injected dose must be adjusted in function of the effective needs in ozone of the water. The simplest way therefore is to obtain a residual ozone concentration at the output of the reactor. Moreover, the reactor to be set in, is generally best chosen after preliminary investigations and tests. The best compromise is to be adopted considering constructional and operational parameters.

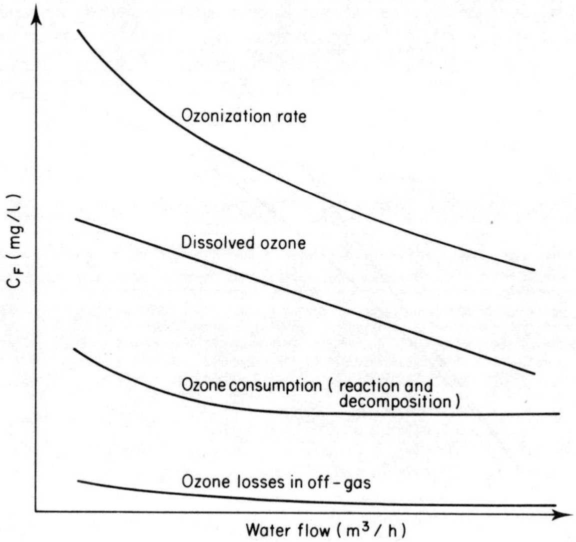

Figure 2. Ozone consumption rates

CHAPTER 27

Contact time of ozonization

J.-P. LEGERON

Trailigaz

When mentioning ozonization, one refers to ozone injection in water. Although the treatment rate is the most known parameter, there is a second factor which is also very important: the contact time.

Thus, a given quantity of ozone can be introduced during a period of time that can vary by a factor of 1 to 5, sometimes more. Likewise, if we consider that oxidation and germ destruction reactions take place during this contact time, the compartment where ozone is introduced plays the role of a chemical reactor. Due consideration being given to the variability of oxidation velocities, it is necessary to

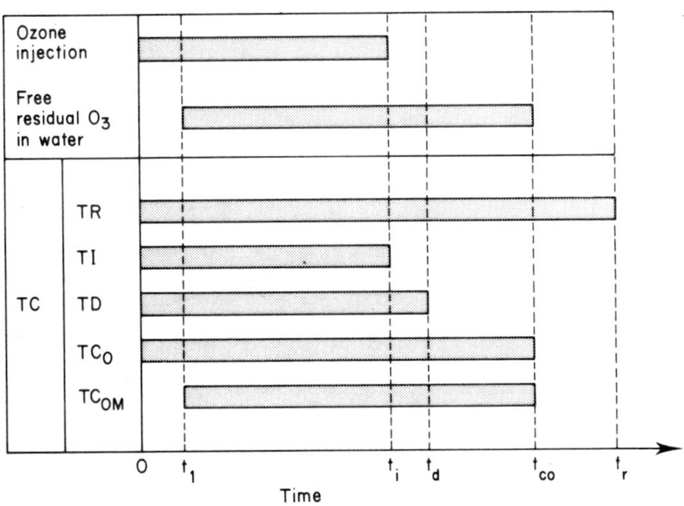

Figure 1. Some meanings of the concept of contact time. TC: Contact time; TR: Residence time (water in a contact chamber); TI: Ozone injection time in water; TD: Ozone dissolution time in water; TC_C: Dissolved ozone contact time in water; TC_{CM}: Measurable dissolved ozone contact time in water

define better the concept of contact time in order to understand the action of ozone on the various components of water.

Figure 1 gives a summarized account of the various interpretations of contact time. For practical reasons, we shall restrict the discussion to injection time (TI) and stay time (TR) of the water in the ozonization compartment.

According to the degree of oxidability and oxidation kinetics, it is convenient to have a more or less long injection time. For instance, if iron is free, a short injection time will be sufficient for oxidizing it (TI $\simeq \frac{1}{2}$ to 1 min). If iron is in a complex form (with humic acids, silica, etc.), a longer injection time is necessary (TI $\simeq$ 2 to 5 min). When disinfection is required, it is always desirable to have a rather long injection time (TI $\geqslant$ 6 min): in addition to the continual transfer of ozone, the presence of ozone gas bubbles induces a range of high concentration ozone gradients in the neighbouring areas.

Ozone can be injected in only the first part of the ozonization compartment whereas the second part is used for water retention. In the case of manganese oxidation, for instance, the second part of the ozonization chamber is used for the reduction of permanganate into manganese dioxide which, in the colloid form, can easily be separated by sand filtration.

Figures 2 and 3 summarize the importance of injection time according to the oxidation kinetics of the organic and mineral matters present in water. Reference has to be made for the place where water is ozonated. Figure 4 shows some types of ozonization chambers. Either pressurized or not, they usually contain several

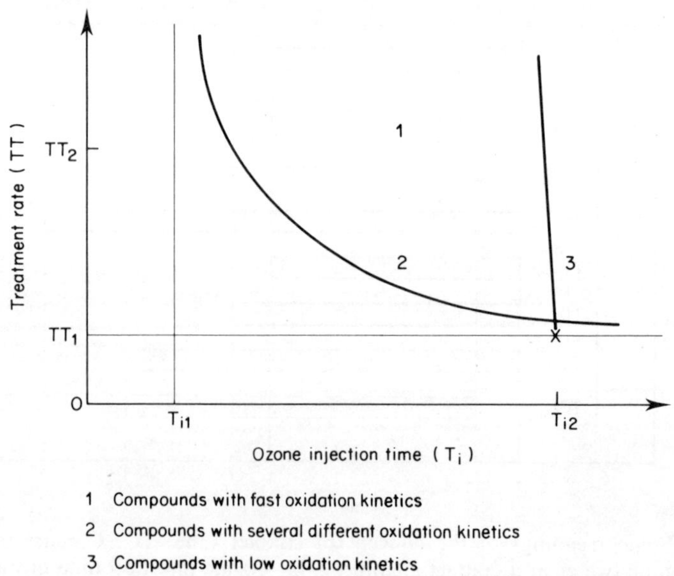

Figure 2. Ozone injection time influence on a 50% reduction * for different compounds (*percentage given as an example)

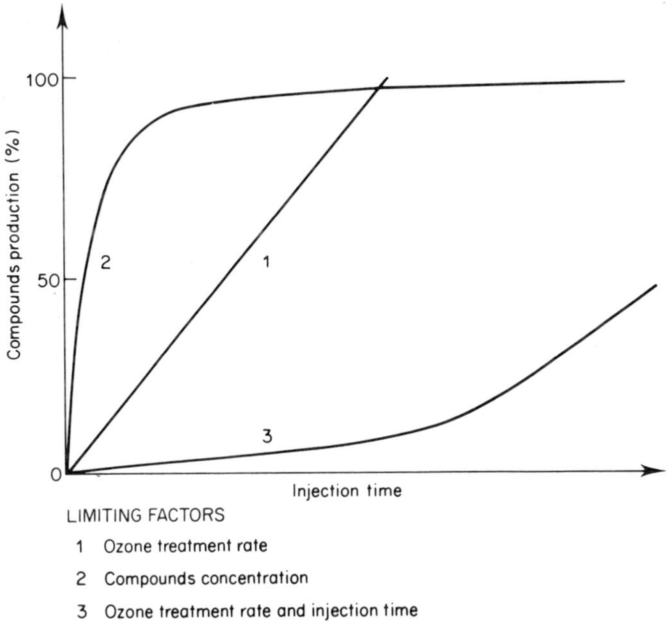

Figure 3. Ozonization kinetics for different compounds with constant ozone flow

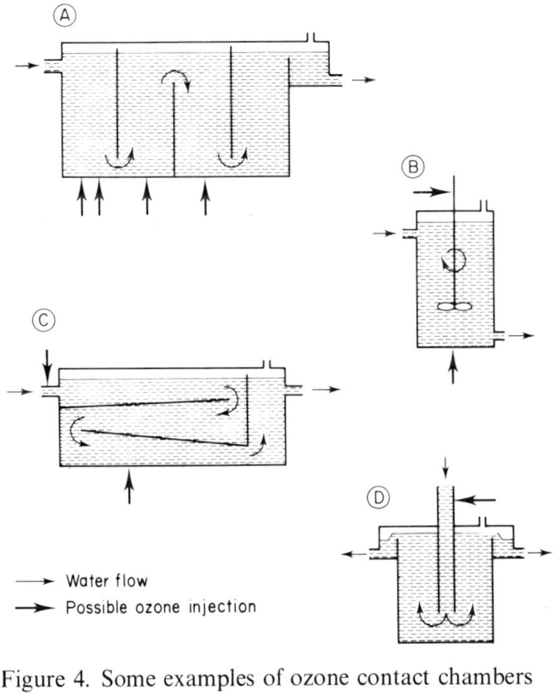

Figure 4. Some examples of ozone contact chambers

compartments where water flows either counter-flow or in parallel with ozonated gas. The chambers are generally made of concrete, but stainless steel can also be used when they have a small size. In short, and although it is impossible to generalize, the contact time depends on the objective contemplated: for preozonization, contact time may vary from 10 s to more than 2 min; for intermediate ozonization, contact time will represent 2 to 5 min. For post ozonization, it will be between 6 and 12 min. However, only tests conducted on each type of water to be treated can optimize ozonization, and more particularly the contact time necessary to reach the objective.

FURTHER READING

J. Chedal, *Diffusion étagée*, International Symposium Eau Ozone, Berlin, May (1977)
J. P. Legeron, *Ozonews*, Part 2, Technical Paper section, **5** (8), 1–6 (1978).
J. P. Legeron, American Institute of Chemical Engineers 86th National Meeting, Houston, 1–5 April (1979).
J. P. Legeron, *l'Eau et l'Industrie*, **34** (April), 74–7 (1979)
W. Masschelein, G. Fransolet and J. Genot, *Water and Sewage Works*, December, pp. 57–8 (1975); January, pp. 34–6 (1976)

CHAPTER 28

Contact columns and bubble-dispersing systems

W. J. MASSCHELEIN

Compagnie Intercommunale Bruxelloise des Eaux

The ozone exchange rate between the gas and the liquid phase is proportional to the external contact surface between gas and liquid. To increase this surface, a principle of dispersion through gas bubbles is generally adopted: either the dispersion of gas bubbles into the water or the spraying of water on contact towers, packed towers or thin-layer liquid contact towers. The major criterion for choice is economical. The spray-contacting systems are generally less expensive in investments than are gas-bubble dispersing. However the solubility of the gas to be dissolved must be high so as to limit the number of gas-transfer stages (equivalent exchange plates). Consequently, direct-spray contact columns are generally not used as in addition the ozone losses are high. Packed columns with Rachig-rings, Berl-saddles or similar material and also plate columns seem interesting because of the increase of their contact surface compared to that of the spray-towers. The 'wet surface' currently ranges between 200 and 250 m^2/m^3.

Therefore when using packed columns the exchange surface in comparison to the column bubble-exchange area requires an equivalent theoretical transfer height of three times less. Whence, constructions may be lower, e.g. 2 m. Packed column exchangers often seem more interesting for the obtention of higher ozone concentrations dissolved in water than what is obtained through bubble-contacting. However in practical circumstances of water treatment, it is generally unnecessary to obtain high dissolved ozone concentrations. Moreover the head-loss in packed columns currently attains 2 m (water height) per metre column height; whence the use of packed columns often leads to substream technique in which part of the total water flow is over-ozonized and subsequently diluted into the main water flow. The substream usually ranges between 8 and 15 % of the total flow while the absolute pressure of the substream attains 2 kg/cm^2. In such circumstances, the costs to dissolve ozone range up to 20–50 Wh/g diffused ozone. Consequently, the method is energy-consuming. Also, taking into account both fundamental and economical aspects of this problem, ozone is most often diffused into water with a bubble-forming equipment, e.g. porous diffusors, turbines, ejectors, etc. Moreover, multiple diffusion is usually applied in consecutive contact chambers.

The contacting principle is that of a plug-flow reactor. One of the essential features of the dispersing systems is to promote the formation of bubbles with a high dynamic exchange surface and no 'rigid' bubbles with less exchange capacity; whence the minimum bubble-radius must be higher than 1 mm.

Essential characteristics for mixing are:

(a) Reynolds number below 2000 and a velocity gradient attaining 150 s^{-1} when mechanical mixing is applied. Costs depend on the practical system but are to be fixed between 2 and 30 Wh/g diffused ozone.

(b) The velocity gradient due to bubble formation and expansion can be computed by the Camp's formula:

$$G\text{avg} = \sqrt{\frac{Q \text{ gas} \times h \times 10^5}{\mu V\left(\frac{\text{h}}{2} + 10.33\right)}}; \text{ and } G_{\max} = g\rho D_b/6\mu$$

h = height of water layer; μ = absolute viscosity (at 20°C = 10^{-3} s m^{-2}); V = volume of the liquid; ρ = density of the liquid; D_b = diameter of the bubbles; $g = 9.81$ m s^{-2}.

Table 1 reports an over-view of the cost–benefit of different contacting systems.

TABLE 1. Comparison of contacting systems

System	Major advantage	Major drawback	Average estimated operation cost (per g O_3)
Dispersion with porous elements	Static	Clogging and channelling	2–3 Wh
Repeated static injection	Static	Critically flow dependent	2–3 Wh
Static mixing	Static	Losses	4–5 Wh
Sonic mixing	Static	Costs	Unknown but high
Total injectors (fast)	High contact rate	High losses	15–20 Wh
Pipe injection (hydrokinetic)	Low cost	Bumping	0–5 Wh
Partial injection high counter pressure			

Table 1. *Continued—*

System	Major advantage	Major drawback	Average estimated operation cost (per g O_3)
low counter pressure	High turbulence dissolution	Partial over- and under-ozonization	10–45 Wh 4 Wh
'Slow downflow' injection	No moving parts	Channelling	4–5 Wh
Spray towers	Low investments	High losses	Unknown
Packed/Plate column	High yields	Clogging and pressure dependence	15–40 Wh
Gas impeller	Construction of low investment	Less ozone–bubble contact (losses)	2–3 Wh
Surface turbine	Accessibility	Losses	7–10 Wh
Vortex system	Simplicity	(Experimental) and unstability	4–6 Wh
Recirculating propeller	High bubble contacting	Moving turbines	5–7 Wh

FURTHER READING

W. J. Masschelein, Contacting in water including off-gas treatment. In *Ozonization Opus*, eds. R. G. Rice and A. Netzer. Ann Arbor. (In press)

W. J. Masschelein, *Belgian Experiences in the Ozonization of Water*. I0A Congress Berlin (1981).

CHAPTER 29

Materials resistant to corrosion and degradation in contact with ozone

F. DAMEZ

Compagnie Générale des Eaux

Ozone is a powerful oxidant and it is necessary to choose materials that can resist it.

In contrast to other reagents which can be stored, ozone is unstable and must be produced on site by means of complex installations. Efforts made at the design stage to use the most resistant materials will ensure steady operation of the facilities and uninterrupted ozone treatment capacity. The supplementary investment costs are largely compensated by the reduction in spare and emergency units and by decreased maintenance expenses.

To ensure personnel safety in ozone facilities, it is necessary to maintain the ozone content of the ambient air in the premises below the standard of 0.1 ppm; that is, to practically nil. In these conditions, a large part of the equipment has no contact with ozone and can be treated conventionally.

In what concerns the equipment in permanent contact with ozone, there are three distinct categories of exposures:

(a) to ozonized dry air: the concentration is 10 to 20 grams of ozone per normal cubic metre of air;
(b) to ozonized moist air (off-gas): about 1 gram of ozone per cubic meter of air;
(c) to ozonized water: between 0.1 and 0.4 grams of ozone per cubic metre of water.

For each case, Table 1 indicates the characteristics of resistance to corrosion and degradation of most of the materials commonly used in industrial installations.

We can conclude that the most secure materials are the following:

METALS

High-quality stainless steel stabilized with molybdenum (type 316L) has an excellent corrosion resistance in both dry and moist environments. It is the best

Table 1

Materials	Ozonization type of exposure			Comments
	Dry air	Moist air	Water	
Metals				
Chromium nickel silver	***	***	***	
Brass	***	***	***	
Aluminium	****	*	*	
Aluminium alloys	****	*	*	
Pig iron	****	****	****	slow corrosion
Galvanized steel	***	**	**	not shock resistant
Stainless steel	****	****	****	without chlorine present
Sintered stainless steel	*	*	*	
Plastics				
PVC supple	***	***	***	
PVC rigid	****	****	****	without pressure
Vinylester resin	****	****	****	
PTFE (Teflon)	****	****	****	
Polyamide (Nylon-rilsan)	—	—	****	
Epoxide (Araldite)	—	*	*	
Synthetic rubbers				
Chlorosulfonated polyethylene (Hypalon)	****	****	****	with appropriate charge
FPM (viton)	***	**	**	with appropriate charge
Silicone	*	**	**	with appropriate charge
Ethylene propylene	****	****	****	with appropriate charge
Polychloroprene (neoprene)	***	**	**	with appropriate charge
Concrete	****	****	****	
Glass and ceramics	****	****	****	

**** long-lasting; *** usable; ** for low ozone concentration; * quickly degraded.

suited material for ozone generators, ozonized air distribution pipes, the metallic parts of injection devices, and residual ozonized air destruction equipment.

Several safety precautions are nevertheless necessary. The better finished the surface of the stainless steel, the better its resistance to corrosion will be. In particular, welding must be done carefully. Sintered steel alters very quickly.

It should be noted that even very high-quality stainless steel may not have a good resistance to attack by acids. Therefore, all precautions must be taken to avoid the presence of humidity in the ozonator, which leads to very corrosive nitric acid formations. Similarly, the presence of chlorine in the contact chambers must be avoided as far as possible. Finally, devices to eliminate the very acid foam which forms during treatment must be installed.

We note that the use of rigid PVC would provide a good resistance to acids. Yet its application for ozone remains in dispute, particularly as concerns ozonized air under pressure.

SYNTHETIC RUBBERS

Our interest lies mainly in the elastic joints used to ensure the tightness of the ozone generators and of the distribution and injection circuits. Hypalon is the most ozone-resistant synthetic rubber; next is ethylene propylene.

CONCRETE

The use of concrete for construction of contact apparatus guarantees very good security, provided the following precautions are taken:

(a) to choose a cement of very low calcium aluminate content, corresponding to the ocean-setting standard;
(b) to place the reinforcements in the concrete at least 4 centimetres from the surface of the moistened wall;
(c) to use aggregates free of granite shales and micas.

CERAMICS

Where injection is done by means of porous diffusers, ceramic stone remains the most recommended material for industrial use.

CHAPTER 30

Corrosion protection in ozonization

G. E. KURZMANN

Ing-Büro Wallorf

GENERAL

Recently ozonization-praxis has suffered from the damages caused by ozone to reservoirs and containers built in steel of the series MU St. 37.2 even though passively protected. The major causes of damage are the following:

(a) inappropriate design and construction;
(b) inadequate preparation of the metallic surfaces before coating;
(c) application of coatings without regard of the state of art;
(d) insufficient layer-thickness of the paint;
(e) too short curing-time;
(f) ignorance of the mechanisms of deterioration.

In the present state of art, organic coatings are to be preferred to protect steel against ozone attack. Furthermore, the manufacturers must clearly state the resistance of rubber coatings, for instance, in the case of ozonization; this is in view of the food-allowance rules according to art. 1, section 5 of the chapters 30 and 31 of the law on foodstuffs and materials in contact with them (as applicable in Germany).

DAMAGE TO THIN- AND THICK-LAYER COATING SYSTEMS

The long-term behaviour of the coating systems in contact with water containing ozone established by numerous observations as a function of the enamel composition, dry-layer thickness and post-treatment influencing the ozone concentration dissolved in water. The deterioration of the passive-protection layers is only a question of months or years, all depends on the preparation of the surface, the product applied and the skill of the application. The most exposed equipments are the ozone contact columns such as packed columns and adsorption filters, e.g. on activated carbon. Moreover, water-temperature above

20°C and the simultaneous presence of active and free chlorine favours water corrosivity. Steel vessels installed in swimming pools are particularly exposed to this problem.

To be considered as the most significant causes of damage through the attack of the coatings by oxidants are: the retention of water, the diffusion of water vapour, oxygen or ozone. In some circumstances if metal is exposed it may be enhanced by local formation of galvanic elements.

The effect on the coating systems of dissolved ozone in water cannot be neglected even at concentrations below 1 g O_3/m^3 as can be observed through the damages caused to activated carbon filters. As a result of the lowering of the concentration due to the deozonization action of granular activated carbon, ($C + O_3 = CO_2 + O$), the attack of the coating system as well as the iron corrosion are both decreased in the direction of the water flow. Furthermore electrochemical measurements have demonstrated that corrosion of St. 37.2 iron in natural water occurs as follows:

Untreated drinking water (T = 10°C) = 0.4 mm/year
Chlorinated water at ca. 0.2 g Cl_2/m^3(T = 28°C) = 0.72 mm/year
Chlorinated and ozonized water ca. 0.3 g Cl_2/m^3 = 1.37 mm/year
and 0.8 g O_3/m^3 (T = 28°C)

Data from the most recent investigation indicate that on multi-phase ozonization at concentrations $> 2\,g\,O_3/m^3$ establishing a redox potential higher than +1,000 mV (Pt-calomel), a corrosion depth of over 2 mm iron per year can occur. Even in hard waters, no protective layer has any significant effect. The attack of the steel occurs according to the well known reaction:

$$2\,Fe + 2\,H_2O + O_2 = 2\,Fe^{2+} + 4\,OH^-$$

CRITERIA FOR COATING APPLICATION

As raw materials become scarcer and requirements for the protection of the environment and safety in working premises increase, a new impulse is given to research in coating and finishing-enamel industries. The specifications for organic coatings applied to protect steel vessels against corrosion must not only have a sufficient mechanical resistance but also be acceptable in respect of the regulations on foodstuffs. In the first instance, a dry-layer thickness of at least 300 μm must be obtained. To reach a more permanent lifetime of the protection system, the following requirements must be met: high resistance to water, high-level vapour diffusion barrier and good ageing quality. To obtain a permanent protection against corrosion a careful and adequate preparation of the surface to be protected is essential. This action implicates not only the removal of rust, carbon residues and welding joints but also the blasting in any area must be efficient and the construction geometry must be designed so as to avoid dead-areas. Furthermore, the application of organic coatings must be carried out according to all guidelines for environmental protection. The application of the corrosion-protection system must be carried out under the control of a professionally qualified applicator and a control procedure foreseen after each layer is applied.

COATING SYSTEMS

Even when all conditions are met for the construction of the tanks, surface conditioning, coating material and its application it still remains difficult to guarantee a permanent protection for several years of exposure to ozone. However long-range protection can, at present, be obtained through rubber-lining but the system is only rarely accepted for ozonization equipment. The reasons therefore appear to be less based on economic arguments, rather than on market-competition arguments, seeing that 'soft coatings' with good behaviour to temperature changes, great elasticity, self-vulcanizing and high resistance to ozone are preferable to hard coatings. Especially in cases of pre-existing corrosion of steel vessels, elastic coatings may be better than the presently favoured hard coatings.

To determine the possible lifetime of an auto-vulcanizing soft rubber coating, the probe is introduced into an ozone contacting column and exposed for 360 h at 14°C to water containing ozone in the concentration range of 0–2.4 g O_3/m^3 and chlorine at 0–2.3 g Cl_2/m^3. Part of the probes can be located for 50 % in the gas zone and exposed there to 20 g O_3/Nm^3. In these extremely severe conditions it has been possible to estimate the theoretical lifetime of soft-rubber coating systems to 7.2 to 10.8 years. In practice, even longer lifetimes can be expected as the attack of the material decreases in intensity within the service time. The materials submitted to the tests up to now are in agreement with the requirements to be fulfilled for contacting and exposure of potable water.

Finally it must be considered that for the application, the steel construction, the surface preparation, the application and the curing require very high attention.

FURTHER READING

Gesetz zur Gesamtreform des Lebensmittelrechts, vom 15 August 1974; *Bundesgesetzblatt*, Teil I, Nr. 95 (1974)

G. E. Kurzmann, Schutzfilter und ihre Bedeutung in der Industriewasserversorgung, *Maschinenmarkt*, **83** (60) (1977)

G. E. Kurzmann, Korrosionsschutz von Stählen bei der Ozonanwendung in der Wasseraufbereitung; *Vortrag OZON-Symposion NIZZA*, 25/26 January (1979)

CHAPTER 31

Safety measures to protect the equipment

F. DAMEZ
Compagnie Générale des Eaux

Ozone treatment involves particularly complex technical implementation. So, to ensure an easy operation of the ozone production and application facilities, specific preventive measures are necessary.

Relative onerous equipment, such as ozonators, can suffer considerable damage in consequence of minor accidents during the air treatment and water cooling processes. It is therefore necessary to furnish this equipment with safety devices to detect dangerous anomalies and to avoid any possible damage.

AIR PREPARATION

The air to be ozonized must first of all be cleared of any trace of moisture, otherwise the ozonators will be damaged by nitric acid formation.

To obtain the necessary dew point, i.e. $-60°C$, the air preparation process must consist of three stages: water cooling, freon cooling, and final drying by adsorption.

The adsorbent material of the dryers will quickly become unusable if it comes into contact with impurities, such as dust or oil. Particular care must therefore be taken of the dust and oil removal systems.

Similarly, an air preparation process must be stopped immediately in the case of a malfunction in the cooling stage, as the absorbent of the desiccator can be damaged if it comes into contact with water.

As to the dryers themselves, the rules for best reliability are as follows:

(a) The quantity of adsorbent is to be provided in proportion to the flow rate: a ratio of 1 ton of desiccant for 1000 m^3/h of air provides a good safety margin. The level of filling must be adjusted as well.

(b) Starting up regeneration: it is advisable to choose a fixed duration rather than a variable duration controlled by the dew point.

(c) Regeneration process: adsorbent products being bad heat conductors, it is advisable for the heating to be carried out by hot air, so as to provide as uniform heating as possible.

Finally, each desiccator must be monitored continuously at the outlet so that any abnormal increase in the dew point can be detected as quickly as possible. Capacitive hygrometers are perfectly suited provided the sensor is positioned at a spot which truly represents the quality of the outlet air.

OZONATORS

To protect the ozonators from an accidental increase in the dew point, it is also essential that there be no air backflow from the contact chambers. To ensure this, the following are required:

(a) a valve at the outlet of the ozonator that closes every time the ozonator stops:
(b) temporary high-rate scavenging with dry air before every start-up;
(c) permanent low flow rate scavenging throughout the stand-by periods.

In the case of air lack, and particularly in the case of water cooling failure, the ozonator must be stopped immediately, so that the tubes are not damaged by overheating. The surest safety measure is to fit the air and water feed circuits with flow failure detectors which automatically switch-off the electrical power supply of the ozonator.

DIFFUSION

It is fundamental that the diffusion yield be monitored to quickly detect any accident of the injection equipment that might lead to deterioration of water quality and leakage of ozone into the atmosphere.

The measurements of the following are needed:

(a) the air flow rates into and out of the contact chambers;
(b) the ozone content of injected air and off-gases;
(c) the residual ozone in the treated water.

AUTOMATION

Safe operation of ozonization facilities requires the ability to detect any abnormal phenomenon by means of monitoring, measuring, and analysing equipment, and the ability to react quickly to avoid any equipment damage.

In view of the complexity of the installation and the need for immediate reaction, many of the safety operations should be performed by closed-circuit automation.

4. ANALYTICAL METHODS FOR OZONE DETERMINATION

CHAPTER 32

Methods for the determination of ozone in concentrated gas mixtures

D. MAIER

Betriebshabil- und Forschungslabor der Bodensee-Wasserversorgung

INTRODUCTION

Three absolute methods, independent of standardization, are at present in use for the determination of the ozone concentration in air or oxygen produced by commercial ozonators: the iodometric method, UV absorption, and the enthalpic measurement. Other methods require a standardization by one of the techniques mentioned, for instance, the chemiluminescence measurement as obtained by reaction of ozone with ethylene; these are not discussed in this contribution.

The three methods described have been controlled for ozone concentrations up to 100 g O_3/Nm^3.

TITRIMETRIC DETERMINATION WITH KI

If the reaction medium is neutral and buffered at pH = 7 the reaction of ozone with potassium iodide develops according to the following equation:

$$O_3 + 2\,KI + H_2O = I_2 + O_2 + 2\,KOH$$

The iodine formed is titrated with sodium thiosulphate:

$$I_2 + 2\,S_2O_3^{2-} \rightarrow 2\,I^- + S_4O_6^{2-}$$

The method enables the determination of ozone in air and oxygen. However other oxidants interfere. The influence of nitrogen oxides can be compensated through specific absorption prior to the analysis or through reaction in caustic soda followed by the determination of nitrates. Chemicals, equipment and materials necessary are: conduits in stainless steel, glass or Teflon, Muenke flasks of 500 ml, three-way valves on Teflon basis, exchange valves in glass with Teflon pin-spindles, membrane pumps resistant to ozone, short connections in polyvinylchloride, barometer, gas flowmeter and timer.

The potassium iodide solution is prepared by dissolving in bi-distilled water 20 g potassium iodide, 7.3 g sodium monohydrogen phosphate (Na_2HPO_4–$2\,H_2O$) and 3.5 g potassium dihydrogen phosphate (KH_2PO_4). A 0.1 N sodium thiosulphate solution is prepared in bi-distilled water. The acidification reagent is diluted sulphuric acid obtained by mixing with care 25 ml concentrated sulphuric acid ($d = 1.84$) into 75 ml bi-distilled water.

To prepare the mixed indicator composed of zinc iodide and starch, 4 g starch is dispersed in an aliquot of distilled water and introduced into a boiling solution of 20 g zinc chloride dissolved in 100 ml of water. This solution is boiled under continuous restoring of the lost water volumes, so as to obtain a clear solution which is diluted while adding 2 g zinc iodide to obtain a 1000 ml solution after filtration. The solution is conserved in a dark bottle and no bluish colour must appear on 50-fold dilution and acidification with diluted sulphuric acid.

Two subsequent Muenke flasks (see Figure 1) are fed with 200 ml potassium iodide solution each and put in line into the gas circuit to be controlled. The three-gate valves are first put in position to lead the gas into the destruction circuit. In this position the gas flow to be measured is set at 1 l/min. The position of the totalizing gas flowmeter is recorded when turning over the three-gate valves so as to admit the gas in the analytical part of the flow scheme. It is not imperative to again measure the time delay during the reaction, as fluctuations up to 20% in the gas flow velocity are hardly significant. The total gas volume passed in the system must be accurately followed and kept between 1 l and 5 l. After the reaction the gas stream is again directed into the destruction circuit and the flow measurement is repeated. The pressure and temperature are checked during the sampling process.

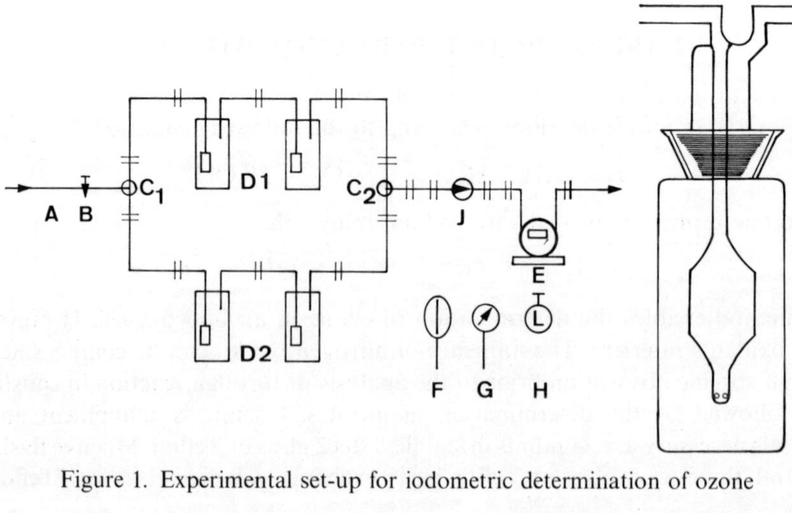

Figure 1. Experimental set-up for iodometric determination of ozone

The iodine formed is titrated with the 0.1 N sodium thiosulphate after acidification with 5 ml diluted sulphuric acid. When approaching the end-point, the yellowish solution is treated with 1 ml of the zinc iodide–starch indicator and further titrated to full discoloration.

PHOTOMETRIC DETERMINATION BY UV MEASUREMENT

Ozone in the range of 200 to 300 nm presents a broad absorption band known as the Hartley band. The maximal absorption index is located at about 253.7 nm. The principle of the method is based on the measurement of the optical density using a monochromatic wavelength of 253.7 nm passing through gas-cell equipped with quartz windows. Perturbing effects can result from all components of the gas mixture absorbing UV light of 253.7 nm wavelength. Such disturbances can be neglected in the usual ozone containing mixtures issuing from ozonators operated with air or oxygen. The measurements themselves are carried out with strict monochromatic UV spectrometers. Important operational parameters to be considered are, e.g., the zero setting to be done when using ozone-free gases which are used to form ozone. Generally a flow of 1 l/min is recommended so as to limit the decomposition of ozone below 1 pm. The molar absorption coefficient is $3024\ cm^{-1}\ mol^{-1}\ l$.

THERMAL DETERMINATION BY ENTHALPY MEASUREMENT

The energy of 144.41 kJ per mol released on the decomposition of ozone represents a measure of the ozone concentration. In its principle the method is based on a precise measurement of the temperature difference between the ozone-containing gas and the reference gas free of ozone obtained when leading the ozone-containing gas on a catalytic decomposing system. The heat given off depends on the ozone concentration in the process gas. Poisoning of the catalyst, as e.g. deactivation through nitrogen oxides, can be avoided with a preabsorbing column for these gases. Moisture must be avoided in the catalytic mass. Other operational parameters are dependent on the commercial version of the apparatus for which guidelines are available.

FURTHER READING

D. Maier and G. E. Kurzmann, *Wasser, Luft und Betrieb*, **21,** 125 (1977)
D. Maier and G. E. Kurzmann, Umweltschutz, *Gesundheitstechnik*, **1,** 8 (1977)
D. Maier and D. Leitzke, *bbr*, **31,** 311 (1978)
D. Maier, *IOA-Proceedings Nice*, **1,** 411, IOA, Paris

CHAPTER 33

Methods for the determination of ozone in diluted off-gases

L. BRENER
Degrémont

Ozone injected in water through various contact methods (porous discs, turbine, etc.) is not totally dissolved. The ozone dissolution rates are up to 95 % and the off-gases contain ozone at concentrations ranging from 0.5 to 10 mg/l of ozone in the presence of oxygen and nitrogen. Such concentrations in ozone cannot be disregarded in view to:

(a) improving ozone dissolution inside the mixing chamber;

(b) anticipating ozone and oxygen recirculation possibilities;

(c) using ozone in preliminary oxidation;

(d) recovering released ozone.

The manual determination of ozone in diluted off-gases proceeds from the same measurement principles as the ozone in gas phase measurement. Its range, however, is reduced—from 0.1 to 10 mg/l—in the presence of a gas mixture with a high oxygen, nitrogen and water-vapour content. The main difficulty of this measurement is the collection of the gas mixture at atmospheric pressure.

The main ozone determination methods use the iodometric determination as a reference.

IODOMETRIC METHOD

The iodometric method permits the determination of ozone at concentrations ranging from 0.1 to 10 mg/l in gas phase. The limit of this method only depends on the volume of gas added to the iodide solution. This determination of ozone is not affected by oxygen and nitrogen contained in the off-gas.

SPECTROPHOTOMETRIC METHOD

The light is largely absorbed by ozone in the wavelength range from 200 to 300 nm, the maximum being 253.5 nm. This property is taken advantage of for the continuous determination of residual ozone by means of a spectrophotometric cell, within the Beer Lambert law range:

$$A = A_0 \varepsilon_\lambda IC$$

where: A = absorbance
ε_λ = molar absorptivity
I = cell path
C = ozone concentration

For measurement at 253.7 nm wavelength, the selected molar absorptivity equals 3024 $cm^{-1}.mole^{-1}$. Low ozone concentrations can be measured with a cell of larger size. Cells with gas circulation are used to prevent the discrepancies of auto-decomposition of ozone.

Ozone concentration is corrected according to the temperature, pressure and water vapour pressure, as per the following equation:

$$C_{O_3} = \frac{48A1000(1 + 0.0037\theta)P_0}{1\varepsilon_\lambda(P - \zeta)}$$

where: C_{O_3} = ozone concentration (0°C—1013 mbars = P_0)
A = measured absorbance
I = cell path
θ = temperature of diluted off-gas, in °C
P = pressure of diluted off-gas
ζ = pressure of the vapour phase at $t°$
ε_λ = molar absorptivity

CALORIMETRIC METHOD

The enthalpy of ozone decomposition into oxygen equals 144.41 kJ/mole. Ozone concentration in a gas mixture can be determined, after passage through an ozone-destroying catalyst, by measuring the temperature difference between the gas mixture with and without ozone. However, this method cannot be applied to residual ozone in diluted off-gases since the water contained in the gas mixture acts on the catalyst efficiency and repeatability.

CHEMILUMINESCENCE METHOD

Ozone reacts with ethylene and emits a fluorescent light intensity which is directly related to the amount of ozone in the gaseous sample.

This method is specific to ozone and, thanks to its sensitivity range (between 0.01 and 10 mg/l), is well suitable for the detection of ozone leaks in the ambient air, all the more that the other compounds which may exist in the air do not interfere with the measurement.

FURTHER READING

G. Gueron, M. Prettre and J. Gueron, *Bull. Soc. Chimique*, **5** (3), 295 (1936)
G. Gueron, M. Prettre and J. Gueron, *Bull. Soc. Chimique*, **5** (3), 1841 (1936)

CHAPTER 34

Monitoring of high ozone concentration in the air

J. VALENTA

Wasserversorgung Zürich

Besides the iodometric titration which is especially suitable for individual measurements, e.g. as a calibration method, O_3-monitors are being installed for the continuous measurements. The majority of these devices work according to a physical–chemical principle, which could possibly simplify the operation and maintenance quite significantly, e.g. when no reagents are necessary.

Together with the requirement of simple operation and sufficient accuracy within the measuring range the following criteria have to be considered:

(a) continuous, direct-reading, recording and reliable measurement;
(b) as high a selectivity for ozone as possible;
(c) as high a zero-point stability as possible;
(d) possibility for connection to a central measuring—respectively a central control system.

The best-known measuring methods for ozone at present are:

(a) the ultraviolet (UV) absorption method;
(b) the caloric method (enthalpy method);
(c) the chemiluminescence method.

UV ABSORPTION METHOD

Ozone absorbs the ultraviolet radiation between 200 and 300 nm (the so-called Hartley bands) with the maximum absorption being at 253.7 nm. This ultraviolet light absorption is proportional to the ozone content and can, therefore, be immediately recorded as an ozone concentration in the gas mixture. The use of quartz cells of various thicknesses enables the application over a relatively wide concentration range.

In customary monitors the mono-chromatization of the UV radiation is attained by means of an interference filter (see Figure 1).

The interrelation between the measured extinction and the unknown ozone concentration is given according to the Lambert Beer law under consideration of the correction for the pressure- and temperature-aberration from the normal conditions.

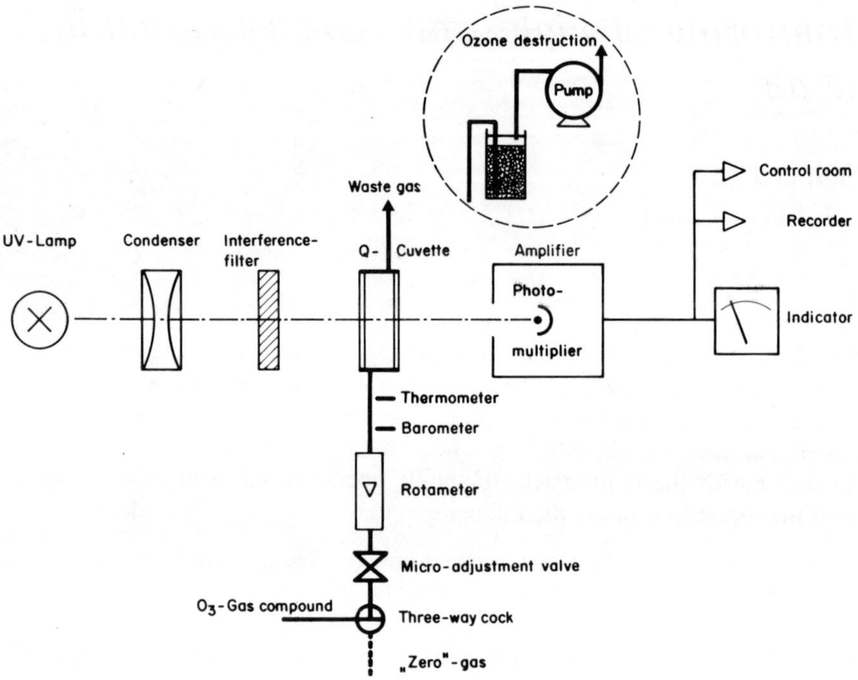

Figure 1. General measuring arrangement of a UV monitor for the determination of the ozone content in a gas mixture

Contamination of the gas mixture under investigation as well as the fouling of the window of the measuring cuvette during a short time are of no consequence because pre-cleaned air (respectively pure oxygen) is used in such a case.

By means of various constructive arrangements in some monitors, other probable sources of errors such as ageing of the mercury lamp and detector, instability of the amplifier and current fluctuation, are mainly compensated, as has been customary for years with most laboratory spectrophotometers. Well-known also is the automatic zero-point correction, as well as the automatic re-calibration.

If the local conditions do not allow the connection of the monitor in by-pass on the process gas flow, the ozone must be destroyed after passing through the measuring system.

CALORIMETRIC OR ENTHALPY METHOD

When an ozone–gas mixture is passed through a catalytic mass in a measuring chamber the ozone decays. The heat released by this process causes a temperature rise in the catalytic agent, namely per mol ozone 144.41 kJ. From the difference between the temperature of the gas mixture at the entry and the temperature of the catalytic agent the ozone concentration in the gas mixture can be deduced.

The nitrogen oxides possibly present in the gas mixture can probably cause a certain inactivation of the catalytic agent. However, this can be prevented with the help of an absorption device.

CHEMILUMINESCENCE METHOD

By this method the intensity of the chemiluminescence, which occurs during the flameless reaction of the ozone with ethylene or rhodamin, is measured. These monitors are mainly employed for lower ozone concentrations (see Chapter 33).

FURTHER READING

DIN 19627, *Ozonerzeugungsanlagen zur Wasseraufbereitung*, Entwurf (1978)

P. E. Erni and L. Pelloni, Bestimmung des spezifischen Energiebedarfes für Ozonerzeuger, *Swiss Chem.*, **10,** 44 (1979)

U. Isacsson and G. Wettermark, Chemiluminescence in analytical chemistry, *Analytica Chim. Acta*, **68,** 339 (1974)

O. Leitzke, Instrumentelle Ozonanlytik in der Wasser- und in der Gasphase; Internationales Ozon-Symposium, *Veroeffentlichung Wasser Berlin*, 164 (1977)

D. Maier and O. Leitzke, Analysenverfahren zur quantitativen Bestimmung von Ozon in Sauerstoff-Ozon-Gasgemischen, *Brunnenbau, bau von Wasserwerken, Rohrleitungsbau*, **9** (1978)

D. Maier, G. E. Kurzmann, Die Bestimmung hoher Ozonkonzentrationen in Luft, *Brunnenbau, bau von Wasserwerken, Rohrleitungsbau*, **10,** 351 (1976)

J. Valenta, Ein Beispel der UV-Kontrolle, *Veroeffentlichung 'Oxidationsverfahren in der Trinkwasseraufbereitung'*, Engler–Bunte Inst., Karlsruhe, 481 (1979)

CHAPTER 35

Determining the presence of ozone in work areas and in the ambient atmosphere

F. DAMEZ

Compagnie Générale des Eaux

The presence of ozone facilities entails a risk that the ozone concentration in the atmosphere may increase, both within the facilities and outside.

Measuring devices are therefore installed to detect any abnormal increase in the ozone concentration in the air, and especially any exceeding the standard of 0.1 ppm, set to protect personnel.

Several types of measurements are necessary, depending on the problem under consideration.

REFERENCE METHOD

First, a very precise reference method is needed to check the other methods, especially those that are operated on a continuous basis. Until now, the method used was the NBKI (neutral buffered potassium iodide) method that was chosen by the ASTM standard D 2912-76. It analyses the iodine released on oxidation in a potassium iodide solution buffered to a pH of 6.8 with a phosphate buffer. A number of recent investigations have shown that this measure gives too high results at very low ozone levels, within the 0.05–1 ppm range. Replacing the phosphate buffer with a borate buffer (BKI method—borate buffered potassium iodide), corrects this defect.

It should be noted that both of these methods are non-specific for ozone and that they measure the overall oxidant content of the air. They should be used only after compensating for the interfering oxidants and are more recommended for reference procedures for calibration than for use as a monitoring means.

MANUAL MONITORING METHOD

Manual methods are practical mainly in two cases:

(a) for routine inspections where installation of continuous analysers is not justified;
(b) during emergency action after leakages, to measure the ozone concentration in the area where the accident occurred.

In both cases the analysis methods must be as quick and as simple as possible. Detection by Draeger tube fulfils these ease-of-use requirements, covering a wide range of measurements, from 0.05 ppm on. Specificity is less critical to be considered in such emergency procedures.

CONTINUOUS METHODS

In the case of fully automated ozone facilities operating with no personnel on site, installation of continuous analysers is necessary. These should be placed in carefully selected locations.

First of all, leak detectors should be put in places which are transit areas for employees and areas where leakage hazards are probable. Electrolyte analysers (e.g. KI, KBr) can serve this function. These simple appliances emit an alarm as soon as the oxidant content of the air exceeds 0.1 ppm. Due to their operating principle, they generally over-estimate ozone levels. This is a minor shortcoming for safety use. As they are relatively inexpensive, it is possible to install a sufficient number of them at a reasonable cost, thus ensuring a localized alarm.

Elsewhere, continuous and precise monitoring of the exhaust to atmosphere is necessary after the destruction of residual ozone at the outlet of the contact chambers. Here, the function of the continuous analyser is to signal any abnormal decline in the efficiency of the destruction equipment as quickly as possible, to avoid the threshold standard from being reached in the area surrounding the ozone facilities. This measurement should be as accurate as possible in the 0.01 to 0.1 ppm range. Two methods are feasible: the chemiluminescent method and the UV adsorption method. The chemiluminescent analysers are based on the reaction of ozone on ethylene. More advisable are UV analysers, which are specific for ozone and require no reagent.

FURTHER READING

American Society for Testing and Materials, Standard test method for oxidant content of the atmosphere (neutral KI), Philadelphia, ASTM; ASTM D 2912-76 (1976)

D. L. Flamm, Analysis of ozone at low concentrations with boric acid buffered KI, *Environ. Sci. Technol.*, **11** (10), 978–83 (1977)

D. Maier, *Standardisierung der Ozonmessung in der Bundersrepublik Deutschland*, European Forum on Ozone Applications to Water Treatment, Nice, 25–26 January 1979, International Ozone Association Committee (ed.), pp. 411–36 (1979)

CHAPTER 36

Determination of residual ozone in water

J. L. HAMON
Compagnie Générale des Eaux

INTRODUCTION

In order to carry out ozonization treatment effectively, operators must dispose of methods for measuring ozone residual in water that are sufficiently sensitive, accurate (concentrations below 1 ppm), and simple (routine use). These analytical methods must be specific and in no case be subject to interference from chlorine or its derivatives.

Further, the growing tendency towards automation in ozone treatment makes automatic continuous measurements increasingly useful, even indispensable.

PRECAUTIONS IN SAMPLING AND ANALYSIS

Quantitative analysis of ozone in water involves several problems related to the gaseous nature of this oxidant, the instability of ozone residuals in water, and the great reactivity of ozone.

In order to avoid significant errors in the measurements it must be ensured:

(a) that the measurement takes place immediately after sample-taking;
(b) that the air to ozonated water contact is limited as far as possible (by avoiding violent agitation);
(c) that the temperature of a sample is not raised;
(d) that the laboratory equipment is very clean and the reagents pure.

MANUAL ANALYTICAL METHODS

Besides the reference method, we will deal here only with the techniques which are, because of their specificity, sensitivity, and facility of use, of greatest interest to water treatment plant operators.

1. Reference method

The iodometric method was long considered the only one usable and is still regarded as the reference method. This method suffers from a lack of accuracy for concentrations below 1 mg/l and from a lack of specificity (interference from chlorine, chlorine dioxide, peroxides, etc.).

2. Methods based on the use of glycine

Glycine has the effect of destroying ozone and of transforming free available chlorine into aminochloracetic acid, which behaves like a chloramine.

(*a*) *Method using iodometry + glycine*

(i) Primary analysis after addition of potassium iodide can determine the total oxidant (ozone + total chlorine).
(ii) Secondary analysis after addition of glycine and then of potassium iodide results in determination of total chlorine.
(iii) Ozone concentration is obtained from the difference (expressed in chlorine).

(*b*) *Method using syringaldazine + glycine*

Syringaldazine is a reagent which is specific to free available chlorine and which does not react with monochloramines or dichloramines: it reacts with other oxidants, such as bromine, chlorine dioxide, iodine, and ozone.

It produces a violet colour which allows colorimetric analysis.

With the method using iodometry + glycine, two determinations of liberated iodine are carried out.

(*c*) *Method using DPD + glycine*

Diethyl-*p*-phenylenediamine (DPD) is also a reagent that is specific to free available chlorine. It reacts with oxidants, such as chlorine dioxide, iodine, ozone, and chlorites and with chloramine it reacts only slowly.

It yields a red-violet product that allows either volumetric titration with ferrous ammonium sulfate or colorimetric analysis.

A process identical to that described above in (a) allows analysis of ozone concentration in the presence of chlorine.

By carrying out several additional analyses, with the DPD method various oxidants (free available chlorine, chloramines, chlorine dioxide, ozone, chlorite) are desdribed as to be determined separately.

3. Colorimetric method using acid chrome violet K

Acid chrome violet K [ACVK = 1,5-*bis*-(4-methylphenylamino-2-sodium sulfonate)-9,10-anthraquinone] is selectively decolorized by ozone or chlorine dioxide. Chlorine, ozonides and peroxides do not interfere.

Interference from chlorine dioxide is a minor concern as chlorine dioxide and ozone are rarely present simultaneously (they react on and destroy one another).

4. Method using indigo carmine

Indigo carmine is selectively decolorized by ozone. The method initially outlined for ozone in air can be adapted for measurement of ozone residual in water (colorimetric and volumetric method). But this method is not accurate at low concentrations (< 1 mg/l). Related methods are in further development.

5. Other methods

(a) The method for measurement of ozone in air using chemiluminescence can be adapted for measurement of low concentrations of ozone in water. This method suffers from several imperfections arising from the complicated manipulations (degasifying).

(b) The method based on the oxidation of Arnold's reagent or of leucomalachite green is specific in regard to other oxidants but requires a complicated operational procedure with chloroform extraction.

AUTOMATIC CONTINUOUS METHODS

Since electrochemical methods or corresponding apparatus are dealt with elsewhere, we will limit ourselves to indicating the various types of electrochemical methods that can be used for the continuous control of ozone dissolved in water:

(a) Specific amperometry using a gold electrode with a counter-potential.
(b) Specific amperometry using a membrane electrode gold/silver cell with a counter-potential.
(c) Specific amperometry using a nickel/silver-silver chloride cell.
(d) Differential amperometry using several cells measuring total oxidant and sequestering reagents.

FURTHER READING

R. N. Kinman, Analysis of ozone: fundamental principles, in Proc. First International Symposium on *Ozone for Water and Wastewater Treatment*, Washington, DC, 2–5 December 1973; R. G. Rice and M. E. Browning (eds), pp. 56–68. International Ozone Institute, Cleveland (1975)

Y. Dorta-Schaeppi and W. D. Treadwell, Uber eine kolorimetrische Bestimmung von Hocheverdünnten Ozon in Sauerstoff, *Helv. Chim. Acta.*, **32,** 356 (1949)

S. Brody, A proposed new analysis for ozone in water using a field portable chemiluminescent ozone analyzer. In: Proc. First International Symposium on *Ozone for Water and Wastewater Treatment*, Washington, 2–5 December 1973, R. G. Rice and M. E. Browning (eds), pp. 84–92. International Ozone Institute, Cleveland (1975)

P. Koppe and A. Muhle, Spezifische Methode zur Bestimmung von Ozon im Wasser in Gegenwart anderer Oxydantien, *Z. Anal. Chemie*, **210,** 241–56 (1965)

L. Dailey and J. Morrow, On stream analysis of ozone residual. In: Proc. First International Symposium on *Ozone for Water and Wastewater Treatment*, Washington, DC, 2–5 December 1973, R. G. Rice and M. E. Browning (eds), pp. 69–83. International Ozone Institute, Cleveland (1975)

J. D. Johnson and J. F. Dunn, Ozone amperometric membrane electrode. In: Proc. Second International Symposium on *Ozone Technology*, Montreal, 11–14 May 1975, R. G. Rice, P. Pichet and M. A. Vincent (eds), pp. 132–42. International Ozone Institute, Syracuse (1976)

W. J. Masschelein, G. Fransolet and R. Goossens, 3rd International Congress of IOA, Paris (1977).

CHAPTER 37

Determination of residual ozone in water with the acid chrome violet K (ACVK) method

W. J. MASSCHELEIN

Compagnie Intercommunale Bruxelloise des Eaux

GENERAL DISCUSSION

Principle

Just like chlorine dioxide, ozone reacts with 1,5-*bis*-(4-methylphenylamino-2-sodium sulfonate)-9,10-anthraquinone (acid chrome violet K: ACVK). The decrease in absorbance at 550 mμ can be measured and related to the residual ozone concentration.

Interferences

Chlorine dioxide reacts with ozone, so that their simultaneous presence and possible interference do not occur in the water treatment practice. The same holds for hydrogen peroxide that by itself does not directly interfere at concentrations up to 10 mg/l.

Chlorine (free or combined) and chlorates do not interfere at concentrations below 10 mg/l. The possible interference of peroxides was tested by using perlauric acid. Unless reaching concentrations of 100 mg/l of perlauric acid, the effect was not significant. Similarly, the 'ozonide' of 2,3-hexene, at concentrations up to 40 mg ozonide per litre, presents no interference; whence it may be emphasized that the method proposed here is not in any way influenced by the possible reaction intermediates brought about by the reaction of ozone with organic compounds at the possible concentration levels in the final stage of drinking-water treatment.

As for manganese, it does not interfere noticeably at concentrations below 1 mg $KMnO_4$/l.

Sensitivity

Using a spectrophotometer and cells of 5 cm optical path, the sensitivity of the method is about 25 $\mu g\ O_3/l$.

Range

The method is applicable in the range of 0–1 mg O_3/l.

Apparatus

Spectrophotometer for use at 550 nm providing a light path of 5 cm or longer.

REAGENTS

ACVK reagent

The product used corresponds to the Color Index code CI 61710 and is at present manufactured by ACNA, a Montedison subsidiary located in Milan (Italy). It is also available at Ugine–Kuhlman in Paris. 175 mg of the dye are suspended in an aliquot of distilled water. To the suspension are added, subsequently, 20 mg of chemically pure sodium hexametaphosphate, 48.5 g of analytical grade ammonium chloride, and 1.6 g of ammonia. The solution is then diluted to 1 l and set aside for 24 h, before use.

This reagent is best conserved in an amber bottle and kept in the cold. It is recommended to renew the solution once a week, especially if exposed to light.

Other reagents

Other conventional reagents are standard sodium thiosulfate 0.01 N, analytical sodium bicarbonate, a phosphate-buffered KI solution: per litre 13.6 g KH_2PO_4, 14.2 g Na_2HPO_4 and 100 g KI, and a starch indicator solution for iodometric determination.

PROCEDURE

Analysis of samples

A 200 ml volumetric flask is filled with 20 ml of the reagent ACVK. Next the water to be investigated is added through a tube, which is dipped into the liquid, so as to reach a total volume of 200 ml. A blank solution is prepared similarly, using water of the same temperature as the sample and exempt of ozone.

Reaction time

The reaction is almost immediate. The absorption intensity is best measured within the 20 min that follow.

Standardization

Solutions of ozone, in either bi-distilled water or distilled water eventually buffered with $NaHCO_3$ and also groundwater exempt of chlorine, chorinated products and organic compounds, are used in the standardization of the method. Comparative determinations are carried out iodometrically and with ACVK. It is of course essential to ensure that the standardization water is free of substances able to interfere in the iodometric analysis. After the standardization, these interferences can be avoided by applying the spectrophotometric determination described here.

Measurements

The absorbance at 550 mμ is recorded by using the dilution water without any reagent for the adjustment of a 100% transmittance. The absorbance of a blank solution of the ACVK without ozone or chlorine dioxide must be measured *simultaneously*.

The sample water and the water for blank dilution are best maintained at the same temperature. The values of A (absorbance of the blank solution minus absorbance of the decolorized sample) vary linearly as a function of the concentration in ozone. The typical standard regression line is:

$$A = 0.007 + (0.22 \pm 0.01)\,\text{mg}\,O_3/\text{l}$$

FURTHER READING

W. J. Masschelein, *An. Chem.*, **38,** 1839 (1966)
W. J. Masschelein and G. Fransolet, *J.A.W.W.A.*, **69,** 461 (1977)

CHAPTER 38

Colorimetric method for the measurement of aqueous ozone based on the decolorization of indigo derivatives

H. BADER and J. HOIGNÉ

Federal Institute for Water Resources and Water Pollution Control

Among the earliest reagents used to measure ozone in gases were indigo and its water-soluble sulfonated compounds. Recently the method has been re-established for the measurement of ozone in industrial exhaust gases by the 'Verein Deutscher Ingenieure' (VDI Method 2468). Recently, it could also be adapted for the analysis of ozone in all types of waters which do not absorb light at 600 nm (blue colour). The method is based on the principle that, at a pH below 3, 1 mole ozone decolorizes 1.0 mole of indigo trisulfonate in an extremely rapid and highly selective reaction. This decolorization of the blue dye can be measured by either visual comparison or by photometry at 600 nm. A detection limit of 10 μg/l ozone for visual comparisons and 3 μg/l for instrumental photometric determinations can easily be obtained.

The characteristics of the method are:

(a) its high sensitivity: the absorbance at 600 nm changes by 20,000 cm^{-1} per mole/l ozone;
(b) no significantly coloured filiated products are formed;
(c) for specificity: peroxides formed by ozonization of natural waters effect additional decolorization slowly, any chlorine present can be masked by additional reagents, and the interferences by oxidized manganeous ions can be corrected for;
(d) experimental methods are adaptable for different concentration ranges.

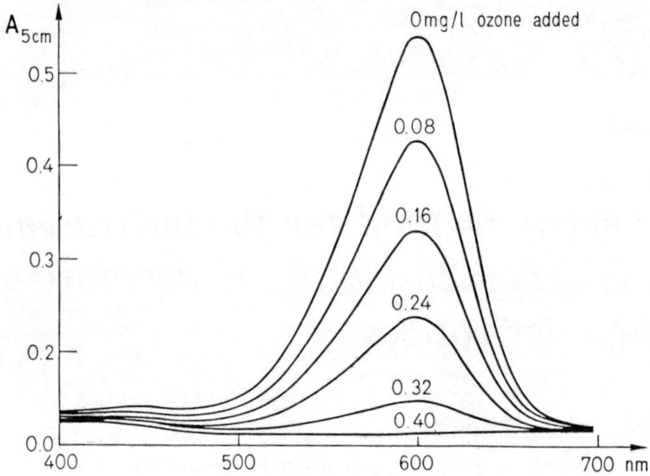

Figure 1

EXPERIMENTAL METHOD

Indigo reagent

Approximately 0.6 g of potassium indigo trisulfonate is dissolved in 1 l phosphoric acid 20 mM. This reagent, when diluted by a factor 100, has an absorbance of 0.16 to 0.18 cm^{-1} at 600 nm. The exact value depends on the purity of the reagent. The solution can be stored up to 4 months or until the native absorbance decreases below 80 % of its initial value.

pH 2 buffer reagent

28 g NaH_2PO_4 (H_2O) and 35 g 85 % H_3PO_4 are dissolved in 1 l distilled water.

Malonic acid reagent

50 g of malonic acid (p.A.) are dissolved in 1 l distilled water.

Glycine reagent

7.5 g glycine (p.A.) are dissolved in 100 ml distilled water.

PROCEDURE FOR OZONE CONCENTRATION GREATER THAN 0.1 mg/l

To each of a series of 100 ml volumetric flasks 1.00 ml of indigo reagent and 10 ml pH 2 buffer reagent is added. For blanks, one or more flasks are filled to the

100 ml mark with water which is free of ozone and other oxidants. Samples of the ozone-containing water are transferred to the remaining test flasks of the series. Volumes of samples containing between 0.01 and 0.04 mg ozone are introduced (pipetting must be carried out with extreme care to avoid losses of ozone, and glassware should be set aside for the exclusive use for this test). After adding the test samples, a slightly blue colour should remain. Totally decolorized zones should not appear for more than a few seconds. Each of the test flasks is now filled to the 100 ml mark with *oxidant-free* water. After mixing the residual absorbance of the ozone containing samples is measured and subtracted from that measured for the blanks. (Absorption measurements can be performed up to approximately 4 h after the samples are mixed.) If the absorbance is measured with a photometer at 600 ± 5 nm, then the slope of the calibration line is 20,000 l/mole . cm) i.e. the change in absorbance is 0.42 per mg ozone added to the final volume as measured. This calibration value is based on an absorption coefficient for aqueous ozone at 258 nm of 2,900 $M^{-1}\ cm^{-1}$. Standard curves have to be determined whenever the measurements are not limited to the wavelength region of maximum absorption (600 ± 5 nm). For calibrations, solutions of ozone in distilled water of low pH can be applied. These solutions must be standardized accurately.

PROCEDURE FOR OZONE CONCENTRATIONS OF LESS THAN 0.1 mg/l

The procedure for low ozone concentrations is essentially the same as the one specified above. However, 10 times less of the indigo reagent should be applied. For this, a secondary stock solution of ten times diluted indigo reagent with shorter lifetime must be prepared. Also, the photometers should be fitted with cells of at least 4 cm optical path-length and with an instrumental error of less than 0.01 units of absorbances in the absorption range below 0.1.

PROCEDURE FOR SAMPLES CONTAINING CHLORINE

Up to 2 mg/l of chlorine can be masked by adding 1 ml of the malonic acid reagent to all of the series of the 100 ml test flasks before the ozone-containing samples are added.

PROCEDURE FOR WATERS CONTAINING MANGANOUS IONS

In this case a differential procedure is necessary. Indigo is also decolorized by oxidized manganese ions. This effect can be corrected by replacing the normal blank by a corrected blank sample in which the ozone was selectively destroyed by reaction with glycine. Specifically the following should be pipetted to two 100 ml flasks (designated A and B): 0.1 ml of glycine solution to flask A and 10 ml of pH 2 buffer reagent and 1 ml of the indigo reagent to flask B. The same amount of ozone-containing samples is added to both of these flasks. After 60 s of reaction time, 10 ml of pH 2 buffer reagent and 1 ml of the indigo reagent are also added to flask A. After adding oxidant-free water to fill the final volume to 100 ml and waiting 30 to 60 min, the residual absorbances at 600 nm are measured in both samples as simultaneously as feasible. The decolorization due to the ozone is found by subtracting the absorbance of the solution of flask B from that of flask A. From this difference the concentration of ozone is deduced such as described above. This procedure cannot be applied in the presence of chlorine.

FURTHER READING

VDI 2468, Verein Deutscher Ingenieure: Messen gasförmiger Immissionen. Messen der Ozonkonzentration, *Manuelles photometrisches Verfaheren Indigosulfonsäure-Verfahren* (1977)

J. Hoigné and H. Bader, *Vom Wasser*, **55,** 261 (1980)

H. Bader and J. Hoigné, *Water Research*, **15,** 449 (1981)

Schweiz. Lebensmittelbuch 27A, *Trinkwasser* (in preparation)

CHAPTER 39

Spectrophotometric determination of residual hydrogen peroxide

W. J. MASSCHELEIN

Compagnie Intercommunale Bruxelloise des Eaux

Hydrogen peroxide may be involved in ozonization praxis and its determination of relevant interest.

GENERAL DISCUSSION

Principle

Hydrogen peroxide added to a concentrated solution of bicarbonate containing cobalt develops a green colour and a UV absorption band with maximum optical density at 260 nm.

Some uncertainty exists regarding the structure of the absorbing compound; the most probable assumption is that of a Co^{III} containing complex carbonate: The cobalt(III)-tricarbonatocobaltate(III):

$$[Co(CO_3)_3]Co$$

This supposed complex has already been applied successfully to determine cobalt but is now proposed in an adapted form for the determination of sub-ppm concentrations of hydrogen peroxide.

The waters are analysed as such. If, however, in very hard precipitative waters turbidity occurs, predilution with distilled water is recommended.

Interferences

The concentrations of nitrate and chlorite causing an absorption equivalent to that of 0.01 ppm H_2O_2, are respectively 100 ppm and 1 ppm.

The complex is stable on filtration and the possible interferences of suspended particles can be eliminated by filtration on 0.45 μm membranes. No interference is noted for chlorates up to 18 mg ClO_3^-/l and NaCl up to 250 mg/l.

Percarbonate and perborate do not interfere but release H_2O_2 on acidification and react according to the quantity of H_2O_2 formed. Persulphates do not interfere in amounts up to 1 g/l and organic hydroperoxides, such as perlauric acid, do not interfere at concentration below 5 mg/l.

Chlorine interferes in the reaction but does not normally coexist with hydrogen peroxide. Chlorine dioxide reacts with the Co(II) reagent but in practical cases ClO_2 is reduced to chlorite by the H_2O_2 (hydrogen peroxide in residual concentrations does not react with the standard ortho-toluidine reagent for residual chlorine).

Minimum detectable concentration

0.01 mg H_2O_2/l can be detected using 10 mm optical quartz cells.

APPARATUS

(a) Spectrophotometer for use at 260 nm providing a light path of 1 cm or longer.
(b) Matched quartz cells of 10 or 20 mm.
(c) Bacteriological filtration assembly for the treatment of turbid samples.

REAGENTS

(a) *Solution of Co(II) at 4 g/l.* Dissolve 19 g analytical grade Co $SO_4 . 7H_2O$ in 1 litre distilled water. When set aside in the dark the solution is stable.
(b) *Bicarbonate solution.* Saturated in distilled water from analytical grade sodium bicarbonate.
(c) *Hexametaphosphate.* Dissolve 10 g analytical $(NaPO_3)_6$ in 1 litre distilled water.
(d) *Standard solution of H_2O_2.* A standard solution of $\pm 510^{-4}$ N H_2O_2 is prepared by diluting analytical H_2O_2 at 30 volumes. The solution is titrated daily by the classical iodometric method.

PROCEDURE

(a) To the 80 ml of a sample placed in a volumetric flask of 100 ml are added successively 1 ml of sodium hexametaphosphate solution, 1 ml of the Co(II) reagent and the mixture is completed to reach 100 ml with a saturated bicarbonate solution (if necessary, the pH of the sample may be adjusted by addition of part of the bicarbonate before the other reagents).
(b) Turbid samples are filtrated on 0.45 μm membranes after the reaction.
(c) The optical density at 260 nm is measured versus a blank solution with 10 or 20 mm cells and recorded.

(d) The blank solution is obtained by treating 100 ml of the water under investigation with 1 g active carbon in powder by magnetic stirring during 10 min. This reference is filtrated on 0.45 μm membranes and seeded with reagents according to the procedure used for the sample.

(e) The optical density is translated in H_2O_2 concentration by interpolating on a standardization curve obtained by introducing known amounts of the standard solution of H_2O_2 into the reference water.

PRECISION AND ACCURACY

The standard error is between 1 and 2% with 1 cm cells in the 0–1 mg H_2O_2/l range. To maintain the reading accuracy at concentrations below 0.1 mg H_2O_2/l the use of cells with 20 mm optical path is recommended.

FURTHER READING

W. J. Masschelein, M. Denis and R. Ledent, Spectrophotometric determination of residual hydrogen peroxide, *Water and Sewage Works*, August, p. 69 (1977)

CHAPTER 40

Operational analysis of residual ozone in treated water

W. J. MASSCHELEIN

Compagnie Intercommunale Bruxelloise des Eaux

Ozone is a gas of limited stability obtained at a relatively dilute concentration in a process gas containing oxygen, e.g. air. No practical case of intermediate storage between its synthesis and application is known. This particular aspect is in contrast to the other reagents generally used in water treatment. As a consequence the operational control and analysis implicates immediate and direct action on the production of the reagent itself and not only on the regulating of the dosing and its equipments.

This very particular aspect also implies a high reliability of the measuring and regulating equipment put in operation in the control of the ozonization and the processes of its production and application.

In the present practice the concepts of automation are based on the regulation of the necessary production to ensure a residual concentration of dissolved ozone, the injection and contacting systems being operated at constant rate.

The analysers in use are based on one of following concepts:

(a) Absorption of UV light at 254 nm. The reliability depends essentially on the geometry and the construction parameters of the cell. These have to avoid the formation of gas bubbles through gas releasing forming gas layers modifying the light absorption.

This problem generally determines the major drawback of such systems of which the application concerns, by its principle itself, only clear water.

(b) Unspecific amperometric determination, with the usual residual chlorine analysers, is often used. The most widespread electrodes are Cu/Au and Pt/Cu. Residual current intensity is inherent to the system.

(c) Specific amperometric analysis is based on the electrode couple Ni/Ag.

(d) Potentiometric analysers are applied through a double cell arrangement comparing the output signal before and after the ozonization of the water.

Whatever the principle, the output signal is the action force for the necessary ozone production. In automation the action operates finally on the voltage applied to the ozonators. Exceptionally the action can also be on the frequency of the current input. As the equipment can be fragile at high or extreme values of the resulting current the necessary devices for safety are:

(a) A delay period in the circuit signal-action.
(b) Limiting of the possible high voltage by a proper automatic regulator so as to avoid the risks of breakages of the dielectric.
(c) Automatic and independent continuous analysis of the ozone concentration in the effluent gas with a reliable apparatus equipped with securities for the maximum allowed concentration level at which the electrical power input is shut off.
(d) The ozonators should be put out of service automatically when the current intensity discharged into the system momentarily exceeds the allowed limits, e.g. 180 Amp for a glass-tube ozonator with a nominal production capacity of 4 kg/h when operated at 20 kV and with air.

FURTHER READING

O. Leitzke, *Proceedings Wasser Berlin*, O. H. Hess, Berlin, p. 164 (1977)
W. J. Masschelein, G. Fransolet, R. Goossens and L. Maes, *Analusis*, **7**, 432 (1979)
W. J. Masschelein, G. Fransolet and R. Goossens, 3ème Congrès International de l'IOA, IOA, Paris, May (1977)
W. J. Masschelein, *Ozonization in Water Treatment*, Ed. Ecochem, Bruxelles
W. J. Masschelein and G. Fransolet, *J.A.W.W.A.*, **69**, 461 (1977)
J. Hoigné, *Vom Wasser*, **55**, 261 (1980)

CHAPTER 41

Principles and design criteria of automation ozonization

J. MIGNOT

Degrémont

BACKGROUND OF AUTOMATION

As ozone is an efficient, but rather costly, oxidant, one is naturally led to control its production and contact with the liquid to be treated automatically to obtain optimum efficiency at minimum cost; this is all the more necessary as large amounts of ozone are used and the price of electric power is increasing. When no automation system exists, operators are faced with many difficulties to reach optimum efficiency at minimum cost.

Only automation permits the following points to be satisfied:

(a) *Dependability of the treatment applied.* For drinking water disinfection, it is essential for the water supplied to permanently have a residual ozone content of 0.4 g/m^3 to obtain bacteriological and viral disinfection. Only a properly designed automation system makes it possible to take all variation factors into account to permanently ensure a fixed residual ozone content and maintain it within strict limits.

Such efficiency can also only be reached by maintaining a high ozone concentration in the air or oxygen serving for ozone production regardless of variations in all of the other factors. How can this be ensured without an automation system?

(b) *Definition of the optimum duty point for the ozonization unit*, resulting in a minimum production cost, in particular when the unit, including several ozonators, need not work at full load.

(c) *Co-ordination of the optimum running conditions for the auxiliary units* used for the compression and drying of the gas to be ozonated, as well as for the injection of ozonated gas.

(d) *Reduction in operating, monitoring and maintenance personnel* when a high-capacity ozonization unit is concerned.

BASIC DATA FOR THE AUTOMATION OF AN OZONIZATION UNIT

1. *The automation of an ozone unit must first fit the mode of action of the ozone*

In certain cases it is only necessary to supply a constant amount of ozone per cubic metre of liquid to be treated; this amount may vary by steps after discontinuous chemical checks. In other cases the automation system must be designed with a view to the results to be obtained after injection, such as the keeping of a minimum residual ozone content during a given minimum contact time, or the absence of an allowance for a minimum limit amount of a compound to be destroyed or oxidized.

2. *Automation depends on selected factors liable to have an effect on the ozonization unit*, both from treatment and energy consumption viewpoints.

The following main factors must be examined:

(a) The variation range of the flowrate of the liquid to be ozonated.
(b) The variation range of the amount of ozone to be supplied per cubic metre of liquid to be treated; along with the 'flow variation' factor, this conditions the design and setting of all subassemblies of the ozonization unit, and in particular the number of ozonators to be put in service and their control system.
(c) The necessity of maintaining a constant ozone concentration in the gas to be ozonized, or maintaining such concentration within a present range. This conditions the number of air blowers to be used and any variation in their rotation speed.
(d) The maximum water content in the gas to be dried and its variation so that it is possible to stop, from time to time, one or several refrigerating units which may be useless or work poorly over an excessively wide range of flowrates.
(e) The interest which may be offered by bypassing one or several dryers in case of a low flowrate of the gas to be ozonated when the water content in the gas is low.
(f) The principle of regeneration of the dryer(s) featuring a variable air or oxygen flow: regeneration carried out as a function of the volume of treated air and the weight of water to be retained.
(g) The oxygen bleed rate in a closed loop with oxygen make-up, as a function of the nitrogen concentration.
(h) The effect of a long contact time at a low flowrate, which may result in ozone losses by auto-decomposition.
(i) The setting of the disposal and/or recirculation system for the residual ozone leaving the contact chambers.

Therefore, the factors to be taken into account must, in each case, be selected after a thorough study has been made.

3. *The automation system can also be designed with a view to a minimum cost per cubic metre of treated liquid*, account being taken of all the elements of the ozonization unit, and not only the ozonators.

4. Automation is only possible when the main factors which condition the running of the unit are known: *it is indispensable to choose and place correctly the monitoring devices* for these factors, as well as their measuring probes and their mode of transmission along with the reading, recording and alarm units which *are best centralized.*

5. *An emergency manual control or servo control* is always necessary, in the event of a failure, for each of the subassemblies of the ozonization unit.

The additional investment cost of an automation system is compensated for by the high flexibility it provides, the accuracy and safety of treatment and the savings made in labour and power consumption.

AUTOMATION SYSTEMS

Various more or less elaborated solutions are available: they depend on the desired degree of automation and optimization.

A relay-type system is sufficient in simple cases: this is the system usually installed in compact or mean capacity units where automation is not complete.

A programmable automation carrying out sequential functions is fitted to more elaborate cases where there are few or no computations to do. It may be completed by a printer and a display console.

A microprocessor-type automation is the ideal solution when computations and programmes are required to optimize the operation of a high-capacity ozonization unit.

In addition to the possibilities offered by a programmable automation, microprocessors integrate the various settings, optimize them *a priori* and *a posteriori*, and make it possible to carry out all desired computations, controls and checks as a function of the experience drawn from start-up, etc.

Practically, computers do not offer larger possibilities. However, when a computer is already scheduled for a high-capacity treatment plant, it can, of course, handle the entire automation of the ozonization unit.

FURTHER READING

P. Fontalirant, M. Pare and J. Guiblais, Un exemple de régulation automatique de la production d'ozone: l'usine de Nantes-La-Roche. 3ème Congrès International de l'IOA, IOA, Paris, 4 May (1977)

M. Marets, Application d'un microprocesseur dans une installation d'ozonation. 3ème Congrès International de l'IOA, Paris, 6 May, IOA, Paris (1977)

M. Pare and J. Guiblais, Ozone automatic regulation. *Proceedings Conférence IAWPR*, London, 17 May, *Pergamon Press*, Oxford (1977)

CHAPTER 42

Analytical instruments for ozone control

Degrémont residual ozone analyzer, Amperazur type

J. GUIBLAIS

Degrémont

The Amperazur unit is an industrial apparatus for continuous measurement of oxidizing agent concentration in the water.

PRINCIPLE

The Amperazur applies a simplified method of analysis by amperometry. The sample cell, continuously fed with the water under analysis, contains a non-consumable, but polarizable gold cathode and a copper anode. When there is no oxidant, the cell thus formed is polarized and a very low residual current circulates. When the water under analysis contains an oxidant, the depolarization of the cell, and hence the intensity of the galvanic current, are proportional to the concentration of the oxidant reduced at the cathode. Such a system allows direct measurement of ozone in water. However, the gold–copper couple of the cell is sensitive to the presence of other oxidizing agents such as chlorine or bromine, which disturbs ozone measurement.

This is why Degrémont has developed an Amperazur apparatus which enables the selective measurement of ozone in the presence of chlorine, which frequently occurs. This instrument uses a couple of nickel–silver electrodes.

However, although it is unaffected by the amount of chlorine, in its construction this cell is sensitive to the nature of the chlorine in the sample. To obviate this effect, that may cause shifts in the correction curves, the eventual presence of chlorine is normalized as chloramines by an injection device for ammonium chloride incorporated in the Amperazur system.

DESCRIPTION (see Figure 1)

The sample cell consists of two plane, circular, parallel electrodes. The flow of sample enters the inter-electrode space tangentially and initiates the rotation of a bed of glass beads designed to favour oxidant diffusion towards the cathode and ensure continuous cleaning of the electrodes.

To feed the cell at a constant rate, the sample is introduced under a constant head (P) providing a flowrate of the order of 60 l/h.

Ammonium chloride is fed at a constant flowrate by means of a 'compressor (Cp)–bubbler (B)' unit designed to apply a constant air pressure, corresponding to the immersion depth of the tube in the bubbler, to the ammonium chloride storage tank (S). A capillary tube (R) on the delivery pipe from the compressor permits the latter's output to be limited.

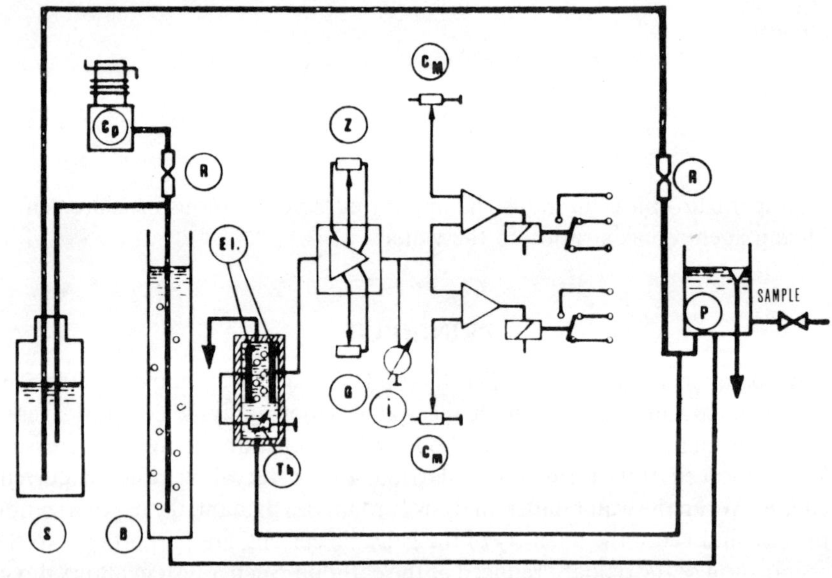

Figure 1

The storage tank is fitted out as a Mariotte bottle. Two air inlet and injection tubes are immersed in the ammonium chloride. The injection tube, fitted with a capillary tube (R), plunges deeper in the liquid than the air inlet tube does. This arrangement allows the capillary tube to be placed under a constant head regardless of the level of the chemical in the tank, and thus enables ammonium chloride to be fed regularly at a low flowrate.

The electrode current of the cell is corrected as a function of the sample's temperature thanks to a thermistor (Th), and next passed into an electronic system where it is amplified to be displayed on a galvanometer (i) as mg/l O_3.

Zero setting (Z) compensates for the residual current, and a second potentiometer (G) adjusts the scale range. The electronic equipment can also send a signal to an external recorder or an ozone production control centre. Furthermore, two comparators permit the sensing of minimum and maximum set points adjustable from 0 to 100% of the scale, by means of two potentiometers (C_m) and (C_M). Whenever the set point is overrun, a changeover switch initiates the control of an external element.

FURTHER READING

W. Masschelein, G. Fransolet and R. Goossens, Mésure de l'ozone residuel dans l'eau—specificité et automatisation. 3rd International Congress of the IOA, Paris, May (1977)

Y. Richard and L. Brener, Mésure et régulation de l'ozone en presence de chlore, European Forum on Ozone Applications to Water Treatment, Nice, January (1979)

In-line testing and display of the content of ozone, chlorine dioxide, chloramines, and free and active potential chlorine in water

A. DERREUMAUX and M. LAMBERT

CIFEC

The CIFEC analyser is capable of differentiating without interference among various oxidizers such as ozone, chlorine, chlorine dioxide and chloramines simultaneously contained in water.

THEORY OF OPERATION

The CIFEC B.S. is composed of several cells for amperometric testing of the total oxidizing equivalent.

Each oxidizer which is to be differentiated corresponds to one cell installed in parallel.

The sample to be tested is fed to each cell. At the inlet of each cell the sample is submitted to a specific chemical masking, except for one reference cell without masking which shows the total oxidizing equivalent; all the other cells show the total oxidizing equivalent minus the specific oxidizer which has been masked.

The differences between the reference cells and all the others are obtained by computer.

The data for each oxidizer are displayed in the front panel of the computer. The electronic system also computes the content in hypochloric acid as a function of the pH and the temperature of the water sample.

APPLICATIONS

(a) Testing of ozone in industrial effluents containing chlorine or chlorine dioxide.
(b) Computation and display of the chlorine demand corresponding to the breakpoint; direct control of an automatic chlorination system at the breakpoint is optional.
(c) In-line testing of free chlorine, combined chlorine, and total chlorine to control the chlorination at the breakpoint.
(d) Differentiation among chlorine dioxide, free chlorine, and combined chlorine.
(e) Automatic operation of dechlorination systems.

DESCRIPTION OF THE EQUIPMENT (see Fig. 2)

CIFEC offers a complete range of analysers with either two, three, or four test cells, depending upon the options selected, which differentiate among two, three, or four oxidizing agents respectively. More cells could be used but in practice the water to be tested rarely contains more than four oxidizers.

The cells are installed on a heavy-duty rack which gives the system a compact appearance.

Each test cell includes:

(a) a motor-driven cleaning device for the electrodes, in continuous operation;
(b) a sample temperature compensating system;
(c) automatic addition of a buffer solution to compensate the variations of pH;
(d) addition of potassium iodide for measuring the total oxidizing equivalent;
(e) addition of specific sequestrant.

The computer is housed with the electronic system in a box away from the cells. The digital displays for the various tests are located on the front panel. All data can be transmitted or used for control by loop feedback.

EQUIPMENT DATA

Power supply	220 V–50 Hz or 110 V–60 Hz
Cell power requirement	10 W
Flow rate of sample water per cell	0.5 l/min
Range	0–1, 0–2, 0–3, 0–5, 0–10, 0–20 ppm

Drift	0
Output current for transmission	4–20 mA with galvanic insulation
Remote recorder	optional

The Anseros analyser for the measurement of ozone in the gas and aqueous phase

K. NONNENMACHER

Anseros

In order to render an economical application of ozone feasible and to produce only as much ozone as is needed, one is increasingly employing continuously and selectively operating ozone monitors which automatically regulate the ozone production analogously or digitally.

The Anseros measuring instrument (Figure 3) works according to the principle of the absorption of electromagnetic radiation. Ozone shows such an absorption at the wavelength of 253.7 nm, thus in the ultraviolet range of the spectrum. In this region the absorption of the energy of the arriving photons always results from a stimulation of the electronic configuration of the molecules.

In the conception of the ozone monitors special consideration was given to the fact that ozone is a strong oxidant which severely attacks any material. We succeeded in making a continuous measuring for process control, e.g. process computers, possible without using parts that are simultaneously being moved mechanically and at the same time come into contact with ozone, such as switch valves. To avoid the fact that drifting of the zero point and soiling of the cuvette have an influence on the precision of the measuring, a special device, to suppress disturbances of any kind even after an extended period of time, was developed.

For the measuring of ozone in the water phase an apparatus for the mass transfer is employed to transform the ozone which has been physically dissolved in water with the aid of injected air into the gas phase and there to measure it photometrically, as has been described above. The special feature of this measuring technique is that it is possible to determine ozone selectively, especially if the water possesses an organically caused absorbance. This measuring technique is used in the treatment of surface waters, such as rivers, lakes, etc., e.g. in flocculation basins, where ozone is used as a flocking agent.

The ozone monitors can be alternatively fitted out for the measuring ranges shown in Table 1. On request the measuring range can also be designed according to the respective area of application.

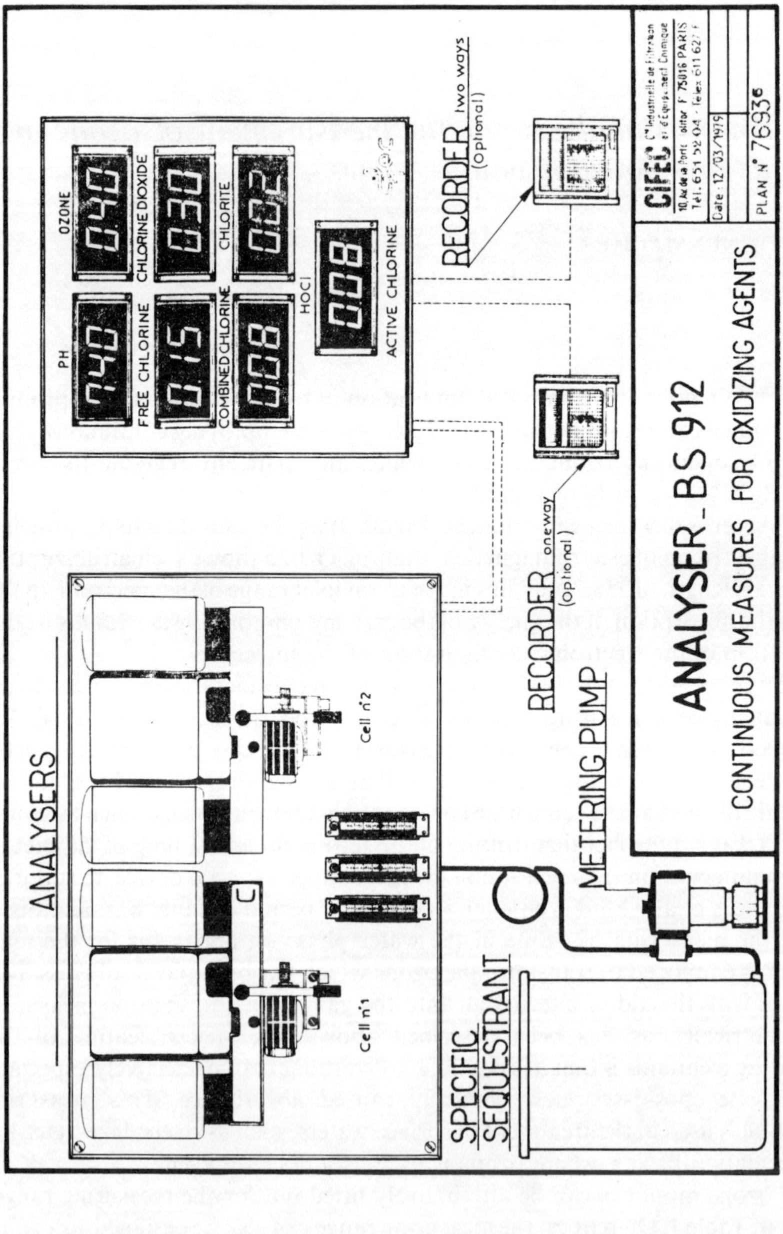

Figure 2. General setting-up of the analyser

Table 1

Measuring range (SI units)	Applications
0–2 mg O_3/m^3 (1 ppm)	Plant safety monitoring
0–10 mg O_3/m^3	Materials testing for ozone damage
0–2 g O_3/m^3	Exhaust air control
0–4 g O_3/m^3	Medical research
0–20 g O_3/m^3	Control of ozone-producing generators
0–40 g O_3/m^3	Dentistry research
0–200 g O_3/m^3	Wastewater treatment

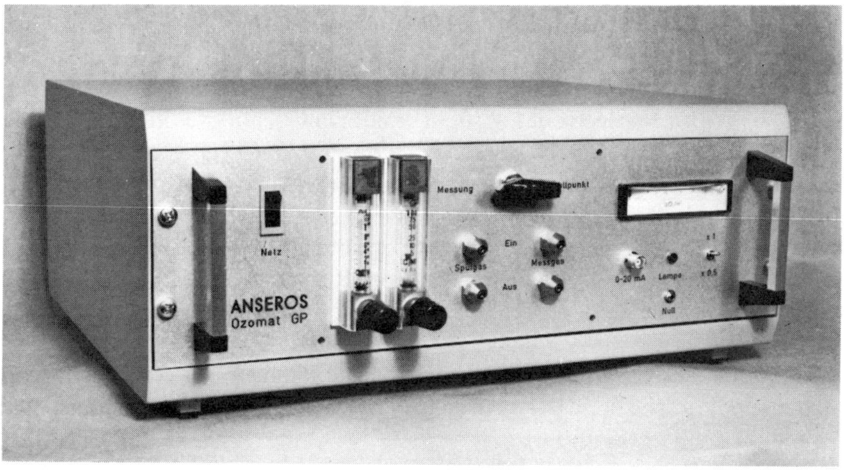

Figure 3

Continuous amperometric residual ozone analysis in the Tailfer plant

W. J. MASSCHELEIN

Compagnie Intercommunale Bruxelloise des Eaux

The original amperometric ozone analyser has been developed for use and operation in the Tailfer plant for the treatment of water from the river Meuse.

The galvanic element is based on the couple Ni–Ag (AgCl). The electrode reactions are assumed as illustrated in Figure 4.

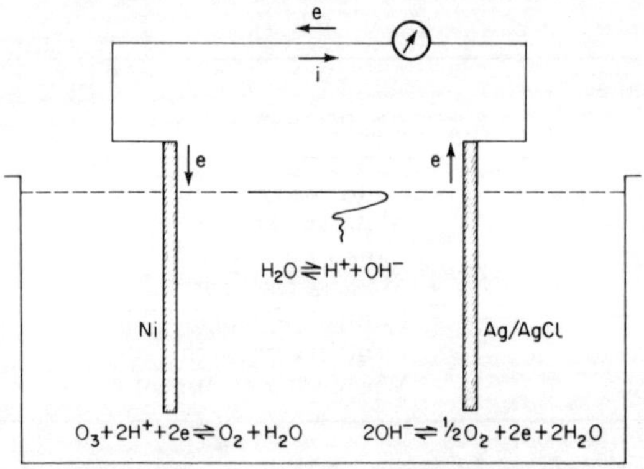

Figure 4. Ni–Ag (AgCl) galvanic element in the presence of dissolved ozone

The main characteristic for the cell is to release no residual current in absence of ozone. With new electrodes, a slight residual current develops with dissolved chlorine but it rapidly drops to zero when the Ag/AgCl cell is constituted. This formation can be accelerated as compared to the action of the chloride ions present in the water by conditioning the electrodes with normal sodium hypochlorite. The depolarization current obtained with ozone is opposite in sign to that of the initial chlorine depolarization. Reverse in the sense of the current when both ozone and chlorine are present is the criterion for good cell conditioning.

The direct-current intensities depend on the detailed cell construction. Our recommendation is to use curved electrodes placed at a distance ranging between 30 and 50 mm (see Figure 5).

The inlet of the water must in its conception avoid transfer of gas (e.g. containing ozone) in the bubble form to the active zone of measurement within the electrodes. Also the distance between the electrodes must be sufficiently high so as to avoid capacitive currents occurring on fouling of the electrode surfaces.

The analyser is based on the ‘acid’ reduction of ozone, whence the galvanic current is lowered in alkaline medium. This is of minor or no importance in the case of post-ozonization in drinking-water treatment but for certain cases of wastewater a pH correction integrated into the measuring circuits can be necessary.

The sensitivity of the system is significantly actuated at a concentration of $0.01\,g\,O_3/m^3$. Maintenance of the electrodes is necessary once a month for standardizing and every 6 months a mechanical cleaning of the Ni electrodes, which are subject to pitting–corrosion.

FURTHER READING

W. J. Masschelein and R. Goossens, *H_2O*, **5**, 100 (1976)
W. J. Masschelein, G. Fransolet, R. Goossens and L. Maes, *Analysis*, **7**, 432 (1979)

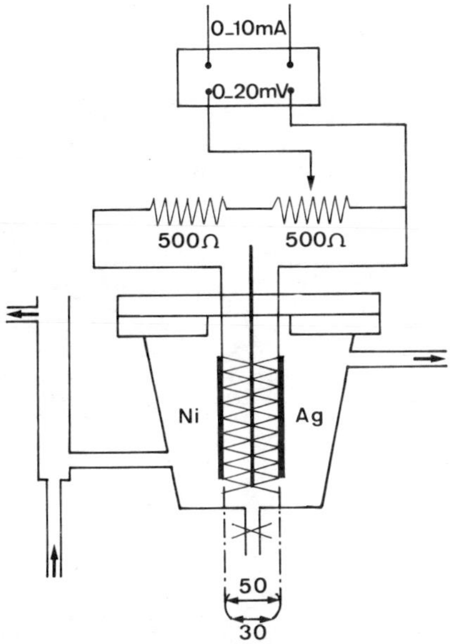

Figure 5. Construction of the Ni–Ag (AgCl) analyzer for residual ozone

The Sigrist A254 photometer for monitoring of ozonators and ozone reactions

W. SIGRIST

Sigrist Photometer AG

In the common ozone–air or ozone–oxygen mixtures, ozone is the only component that absorbs UV light in the 240–300 nm wavelength range. According to German Standard Method for monochromatic light, the molar extinction coefficient (ε) for ozone is 3024 at the wavelength 254.7 nm. The reading of the Sigrist absorption metre model A254 (Figure 6) corresponds directly to the extinction (E) of the sample at the wavelength 254.7 nm. Thus the ozone concentration is:

$$C = \frac{E}{d\varepsilon} = \frac{E}{d\,.\,3024}\,\text{mol/l or } \frac{15.8\,.\,E}{d}\,\text{g/m}^3$$

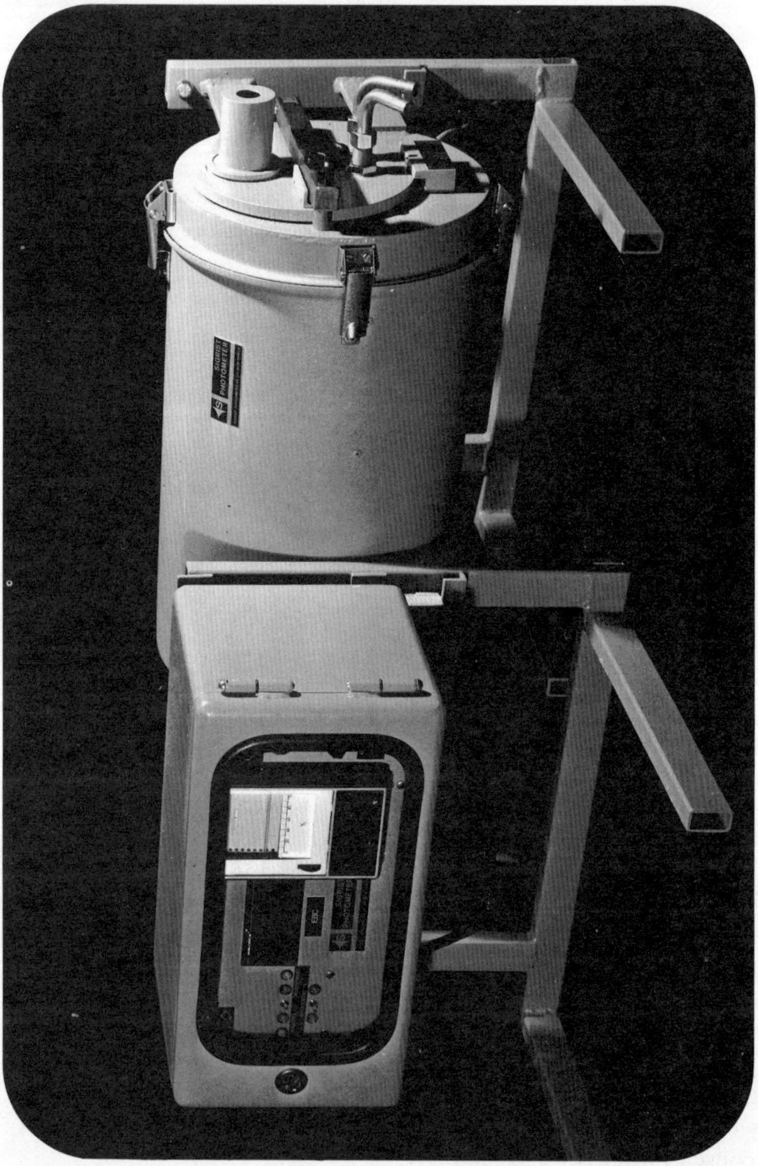

Figure 6. The Sigrist photometers

An ozone measurement based on this principle can serve as an absolute measurement providing the instrument reading represents an unambiguous extinction reading, and providing the extinction reading is determined for monochromatic light at the wavelength 254.7 nm. Both conditions are fulfilled by the Sigrist photometer model A254.

MEASURING TECHNIQUE

The Sigrist measuring technique is particularly suitable for instruments using glow-discharge lamps as light source because the unavoidable fluctuations of the light source have virtually no effect on the readings with this technique. The technique used in the Sigrist absorption meter is represented in Figure 7. A

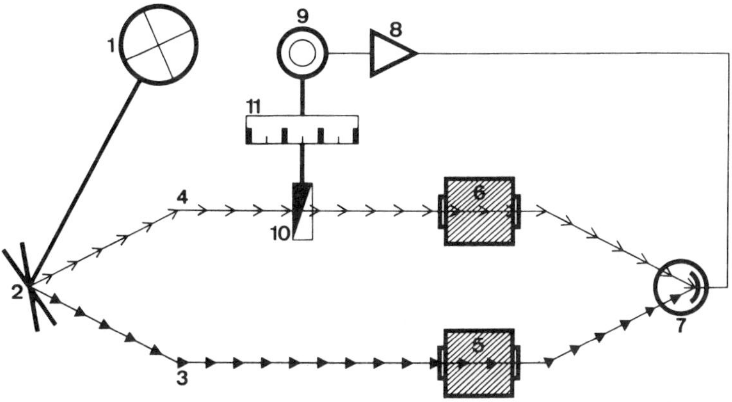

Figure 7. Optical pathway of the A254 photometer

mains-operated light source (1) emits a light-beam on to the oscillating mirror (2) which produces alternately, at about 600 times a second, a measuring beam (3) and a comparison beam (4). The measuring beam (3) enters the flow cell (5), is weakened by the sample flowing through with unknown concentration, and reaches the photocell (7). The comparison beam (4) also reaches the photocell (7) either in its original intensity or weakened by a comparison medium (6). Thus the photocell receives alternately a measuring beam (3) and a comparison beam (4) of varying intensity. It converts the differing light intensities of the two beams into photoelectric current which is subsequently amplified (8) to drive a synchronous motor acting as a servomotor (9). The motor is connected to a mechanical measuring shutter (10) which varies the intensity of the comparison beam (4) until the two beams strike the photocell with equal intensity. The measuring shutter (10) is directly coupled to the measuring drum (11), and therefore the measurement value is a product of the shutter position and can be read on the measuring drum (11). The position of the shutter (10) is altered again only when the concentration fluctuations of the specimen affect the brightness of the measuring beam. This concept underlies the exceptional stability (freedom from drift) of the Sigrist photometer. Worthy of special note is its insensitivity with respect to mains voltage fluctuations, ageing phenomena of the components, and external light.

TECHNICAL DATA

Measuring principle	absorption measurement at 254.7 nm
Measuring technique	double-beam system with optical compensation
Zero point stability	better than ±2% over months
Accuracy	±2%
Response time	max. 15 s
Measurement scope	0.003–190 g/m^3

SERVICE DATA

Power supply	220 V/50 Hz 110 V/60 Hz other voltages on request
Power consumption	44 W (not including flow cell heating; this can be as high as 40 W depending on size)
Ambient temperature	−10°C to +50°C
Explosion protection	connections available for external ventilation
Mounting	wall mounting
Weight	70–90 kg
Dimensions	height 600 mm depth 300 mm width 600–1130 mm (depending on layer thickness)
Medium temperature	up to 50°C for glass flow cells up to 80°C for V4A flow cells (higher temperatures on request)
Pressure	up to 0.1 atm g for glass flow cells up to 2.5 atm g for V4A flow cells (higher pressures on request)
Flow rate	no restriction; approx. $\frac{1}{2}$ l/min is recommended

SPECIFICATION

Basic unit UP62B6

— Optical system of mirrors reflecting the UV light
— Fully transistorized zero amplifier

— Light source B6 (mercury vapour lamp), amperage 30 mA (low pressure), precentred to lamp base
— Steel enclosure coated with polyurethane
— Instrument incorporated in distortion-free frame
— Doors with Plexiglas windows, rubber seals and square bolt closures.

Measuring attachment 5A.OOHHF4

— Interference filter 254 nm
— Photocell F4 (secondary electron multiplier), blue-sensitive, spectral range 200–600 nm
— Heating resistance W, to prevent condensation if medium temperature is below ambient temperature (type 5A.OOHHWF4)
— Dry cartridge D, for dehumidification of the flow cell chamber (type 5A.OOHHWDF4)
— With hermetic flow cell housing).

Flow cells PA 0.5–500 mm

— Material: V4A or glass
— Layer thicknesses 0.5 to 500 mm depending on concentration.

BUILT-IN AND AUXILIARY EQUIPMENT

Examples: direct recorder, control switches, telemeter units, etc., as per separate specifications.

5. CRITICAL QUESTIONS IN OZONIZATION PRACTICE

CHAPTER 43

Toxicity of ozone

F. ERB

Faculté de Pharmacie de Lille

The presence of ozone in our environment is considered either a good thing or, on the contrary, undesirable, depending on the case. Thus, in high altitudes we seek to preserve the ozone layer which allows us to tolerate solar ultraviolet, radiation without danger. On the other hand, in the air we breathe, ozone concentrations as low as 0.3–0.5 ppm provoke detrimental effects on the respiratory tract and ocular mucosa, entailing in the case of acute toxicity (above 1 ppm) the appearance of pulmonary oedema. The concentration in the atmosphere at which ozone is dangerous is not well known but inhalation in an atmosphere of 50 ppm during half an hour would probably be fatal.

Inhalation of lower concentrations, under 0.2 ppm for 3 hours or more, can already result in a considerable dimunition of nocturnal vision, together with an extreme fatigue and an incapacity to coordinate movement. This is why the mean hourly standards (to be surpassed no more than once a year) recommended in the USA are of 0.1 ppm or 0.2 mg/m^3 of air. Canada has adopted the following standards: 0.10 ppm during 1 hour, 0.03 ppm during 24 hours.

The value of purifying the ambient air in a confined atmosphere by using ozone is sometimes called into question. In a general way, what is suspect is ozone's intense reactivity which proceeds by radical or electrophilic mechanisms. It is liable to lead to the formation of ozonides, epoxides, peroxides, etc., and free radicals. These last are considered alkalifying agents, playing a role in chains of deoxyribonucleic and ribonucleic acids (DNA and RNA).

OZONE IN WATER PURIFICATION

It is judicious to examine the effects of ozone on the organic and mineral compounds in solution in treated waters before affirming that its use does not present any health risks.

Action on mineral salts

Ozonolysis permits the destruction of compounds formed by such elements as manganese and iron with organic matter in the water, in particular humic acids. It frees and oxidizes these metals into Fe^{3+} and Mn^{4+} ions, which precipitate if the pH conditions are favourable. Nevertheless, this liberation of various metallic elements can leave traces of metals remaining in the last phase of water treatment (arsenic, nickel, cobalt, copper, cadmium, etc.).

Action on organic matter

If the sterilizing action of ozone appears after 10 minutes contact with water being treated, the degradation of organic matter in the conditions of wastewater treatment is clear. But the nature of the compounds formed and the kinetics of their formation have only been determined in small part. Numerous works are therefore appearing in the literature attempting to elucidate the development of organic compounds under the effects of ozonization. Depending on the case, it appears that the nature of the intermediate products of ozonization can be stated. Thus, phenol leads to maleic acid, the final products of oxidation being glyoxalic acid, glyoxal and oxalic acid, etc. But most of the time, because of the multiplicity and the complexity of the products formed and the unknowns which remain, the toxicity of the products of ozonization is merely estimated.

Toxicity studies

Most authors refer to the mutagenesis tests carried out *in vitro* according to Ames, on bacteria: *Salmonella typhimurium* and on yeast: *Saccharomyces cerevisiae*. They generally use a system of metabolic activation.

A considerable number of compounds found to be carcinogenic using other techniques have been shown to be mutagenic *in vitro* but there are still exceptions such as chloroform and carbon tetrachloride (haloforms).

It seems that nitrogenous compounds have a more apparent propensity to mutagenicity (diphenylhydrazine, nitriloacetic acid, dinitrotoluene, etc.) but the noted activities remain inconstant, low, and out of proportion with the concentrations tested.

Ozonization of dimethylhydrazine, 2-hydroxyethylhydrazine and benzidine furnishes mutagenic compounds of varying stability. It seems that they can be destroyed by prolonged ozonization. On the other hand, ozonization in short periods (30 seconds) leads to the disappearance of the mutagenic character of pesticides such as Captane, Dexon, alkalifying agents such as bischloroethylamine, sodium azide, and aflatoxin B_1 (30 seconds), etc.

Polyaromatic amines: acriflavin, proflavin, β-naphtylamine lose their mutagenic character through ozonization, for 1–4 minutes. Aromatic hydrocarbons are sensitive to ozonization: benzo(a)pyrene, 3-methylcholanthrene, 7,12-dimethylbenzanthracene. Tests before and after ozonization show a marked reduction in tumours, except for dimethylbenzanthracene.

It is hard to extrapolate these results for the treatment of drinking waters and wastewaters because of the infinite diversity of their composition, without mentioning the phenomena of synergy, of potentiation, or of the inhibition of biological effects which cannot be taken into consideration during individual study of pollutants.

This explains the value of total toxicity studies carried out before and after ozonization of the water.

A mutagenic effect appearing after 15 minutes has been evoked. Because of the nature of toxic products, the formation of malonaldehyde and maybe of *N*-hydroxypurines is possible.

A wastewater recycling system for the needs of a mobile surgical hospital using ultrafiltration, reverse osmosis and ozonization resulted in ozonized extracts which can be more toxic than the raw dialysates obtained after osmosis.

CONCLUSION

Finally, the risks most often brought up concern the formation of mutagenic compounds from numerous residual products from industry. The long-term effects of ozonization cannot be foreseen because of the diversity of organic pollutants in traces in water at the end of physicochemical and biological treatment. Furthermore, questions may critically depend on the operational conditions of ozonization as practised: e.g. sufficient dosage and reaction time.

FURTHER READING

B. N. Ames, F. D. Lee and W. E. Durston, *Proc. Nat. Acad. Sci. USA*, **70,** 782 (1973)
B. N. Ames, W. E. Durston, E. Yamasaki and F. D. Lee, *Proc. Nat. Acad. Sci. USA*, **70,** 2281 (1973)
G. R. Burleson, M. J. Caulfield and M. Pollard, *Cancer Res.*, **39,** 2149 (1979)
J. McCann, E. Choi, E. Yamasaki and B. N. Ames, *Proc. Nat. Acad. Sci. USA*, **72,** 5135 (1975)
M. J. Caulfield, G. R. Burleson and M. Pollard, *Cancer Res.*, **39,** 2155 (1979)
J. A. Cotruvo, V. F. Simmon and R. J. Spanggord, *Ann. New York Acad. Sci.*, **298,** 124 (1977)
T. E. Cody and R. N. Kinman, *J.W.P.C.F.*, 1727 (1978)
N. Gruener, *Bull. Env. Cont. Toxicol.*, **20,** 522 (1978)
C. G. Hewes *et al.*, Houston Res., Inc., Texas (1974)
R. W. Ward and G. M. Degraeve, *J.W.P.C.F.*, **46** (1978)

CHAPTER 44

Professional risks associated with ozone

F. VAN HOOF
Antwerpse Waterwerken N.V.

Since the toxicity of ozone to human beings is mainly focused on the respiratory tract it is necessary to pay attention to the effects of ozone on industrially exposed individuals. Among these are: welders, aircraft crew, persons occupied near ozone generating installations or UV sources, etc.

No fatal cases of occupational ozone exposures have been described, but several authors have reported cases of severe ozone intoxication, most of them among welders using inert gas shielded consumable electrodes, or in people working in plants producing hydrogen peroxide. Symptons of acute poisoning include headache, dizziness, burning sensation in eyes and throat, acrid taste and smell, choking sensation, cough and substernal pressure. In some cases, a decrease in visual acuity was observed. Such symptoms only occur after exposure to concentrations higher than 0.3 ppm ozone. Experimental and epidemiological studies confirm these results: no differences in lung function occur below 0.3 ppm. Respiratory ailment disappears within 2 weeks after exposure but this does not necessarily imply complete recovery: weakness, fatiguability and dyspnoea may continue.

Long-term exposure to ozone in occupational situations can lead to manifestations of chronic toxicity: headache, weakness, increased muscular excitability, decreased memory, increased prevalence of bronchitis, emphysema and fibrosis. In some of the studies, however, it is not clear whether the effects are due to ozone or a combination of ozone and other pollutants. It should further be emphasized that ozone intoxication can be overlooked since symptoms are similar to those caused by some common respiratory diseases.

Except for the phenomena mentioned, some aspects of ozone toxicity should be kept in mind when evaluating occupational ozone exposure. Ozone is often inhaled in combination with other pollutants, such as sulphur dioxide and nitrogen dioxide. The effects of ozone and nitrogen dioxide seem to be additive while combination of ozone and sulphur dioxide gives synergistic effects.

Consideration should also be given to the effect of heat and physical exercise: both have been proved to increase the effects of ozone toxicity.

Some reports have mentioned the influence of low levels of ozone on intellectual and athletic performance; however results in this field are scarce and need to be confirmed.

At the actual state of knowledge it is difficult to set a threshold below which these phenomena would not occur. Ozone has been proved to decrease resistance against infection of the respiratory system by different types of bacteria even at levels below 0.1 ppm. Synergistic action of nitrogen dioxide has been described in this respect.

In vitro experiments have shown ozone to have mutagenic properties. These results were confirmed through *in vivo* experiments using mice, indicating a much higher incidence of chromosome aberrations in lymphocytes after an exposure of 40 hours to 0.1 ppm ozone than through exposure over the same period to the permitted dose of radiation. These effects were not entirely confirmed in man after exposure to 0.5 ppm ozone during 5–10 hours. The effects produced on human lymphocytes were nevertheless thought to be sufficient to produce persistent chromosomal aberrations.

No specific treatment exists for ozone poisoning. Symptomatic treatment will include rest, oxygen, cough-suppressants, analgesics and antibiotics.

Prevention of occupational ozone intoxication can be reached by avoiding exposure to people who regularly suffer from infections, diseases of the respiratory tract or are subject to other lung diseases. All people who will be exposed to ozone should be subjected to a preplacement examination. Several countries have set a TLV value of 0.1 ppm ozone for a 40 hours (5 days) per week exposure period.

FURTHER READING

P. J. R. Challen *et al.*, *Brit. J. Ind. Med.*, **15,** 276 (1958)
R. Ehrlich *et al.*, *Environ. Res.*, **14,** 223 (1977)
L. J. Folinsbee *et al.*, *J. Appl. Physiol.*, **43,** 409 (1977)
M. Hazucha *et al.*, *Arch. Environ. Health*, **27,** 183 (1973)
M. Kleinfeld *et al.*, *Arch. Ind. Health*, **15,** 27 (1957)
O. F. Kudrjavceva, *Gig. truda i Prof. Zabol.*, **6,** 52 (1963)
T. Merz *et al.*, *Mutation Research*, **31,** 299 (1975)
A. Nasr, *Clin. Toxicol.*, **4,** 461 (1971)
V. A. Newill, Proc. First. Int. Symp. on Ozone for Water and Wastewater Treatment, IOI (USA) (1975)
A. I. Nevskaja and T. A. Kocetkova, *Gig. truda i Prof. Zabol.*, **1,** 20 (1961)
C. Nogues, *Proc. Third IOI Congress, Paris, IOA, Paris* (1977)
F. Van Hoof, *Proc. Third IOI Congress, Paris, IOA, Paris* (1977)
W. A. Young *et al.*, *Arch. Environ. Health*, **7,** 337 (1963)
R. E. Zelac, *Environ. Res.*, **4,** 325 (1971)
World Health Organisation, Environmental Health Criteria 7: *Photochemical Oxidants*, Geneva (1978)

CHAPTER 45

Protective equipment for emergency action

M. Rapinat
Compagnie Générale des Eaux

Beyond a certain concentration ozone is a toxic gas. Exposure standards have been established in most countries using ozone:

(a) the threshold limit value for an 8-hour work period has been set at 0.1 ppm, or 0.2 mg/m^3;
(b) the maximum concentration for an exposure time of 10 minutes is 0.3 ppm according to the ANSI/ASTM standard E 591–77.

Consequently, during normal operating conditions, the ozone concentration in the facilities and their surroundings must be maintained below 0.1 ppm. This requires the implementation of means for the destruction of residual ozone and for venting. Furthermore, in the case of an accident, means of individual protection are required.

For low ozone concentrations, up to 5 ppm, the use of a cartridge gas mask is generally sufficient. Masks used should consist of ozone-resistant material, conceived on a hypalon base. They should completely cover the face, so as to protect the eyes from the irritating action of ozone. Filter cartridges should be models designed for ozone. As a precaution, the use of cartridge masks should be limited to short-term use, of about $\frac{1}{2}$ hour.

For concentrations above 5 ppm or for long-term use, self-contained breathing equipment is necessary, where the mask is fed with compressed air from bottles or other sources.

Choice of protective means should be made by the person in charge of security, who should accompany the emergency unit and take detailed analytical measurements of ozone in the affected area.

Masks should be allocated on an individual basis. Only compressed air equipment can be of generalized use. In this case it should be checked systematically at fixed intervals and after every use.

Demonstration and training sessions to inform the staff of the proper use of the masks, conditions in which they must be worn, and maintenance, should be foreseen. So as to maintain the efficiency of this instruction, demonstration sessions should be renewed periodically.

External personnel working on the plant or on the equipment should be informed of the risks. Prior to intervention, the person in charge of security should advise concerned firms as far as security regulations are concerned and check that these measures are followed.

In case of accidental inhalation, the victim will have difficulty breathing, more or less so depending on the concentration and the time of exposure. The first measure to be taken is to immediately lie the victim down, so as to reduce the sensation of suffocation caused by irritation of the respiratory tract, while waiting to administer the proper help.

CHAPTER 46

Excess ozone disposal

C. COSTE
Degrémont

The maximum ozone content in the ambience of a working environment has been set at 0.1 ppm by volume. Although higher concentrations can be admitted for short periods of time, the inside or outside ambience of a water treatment plant where ozone is used must necessarily have an ozone content equal to or less than the threshold limit value. As the systems designed for ozone–water contact have a yield of dissolution lower than 1, the plants are led to discharge to the atmosphere air containing varying amounts of ozone depending on the cases, but with a concentration far above the maximum value. In such conditions, besides the risks of toxicity to the human organism and the immediate environment, the electric, electronic and optical instruments and various pieces of equipment in water treatment plants can be subject to irreversible damage.

All these considerations lead to the conclusion that excess ozone disposal is recommended at plants utilizing ozone, more especially in urban areas, before air is discharged to the atmosphere.

DISPOSAL METHODS

In the following, excess ozone recycling will not be dealt with, as this cannot be considered a disposal means in itself: although the efficiency of such a system cannot be underestimated, the disposal of residual ozone after recycling remains indispensable, as the residual ozone content is still far above the threshold limit value after this operation.

Thus, three practical effective disposal methods are at present available:

(a) decomposition by heat;
(b) decomposition by heat with energy recovery;
(c) catalytic decomposition.

The choice of either process can be made based on the following criteria, the list of which is not exhaustive as given hereafter:

(a) available space or ground space requirements;
(b) nature of the existing or required infrastructure;
(c) any incidence of energy consumption;
(d) possible capital cost;
(e) incidence of the amortization rate of the capital investments;
(f) nature of the carrier gas: air or oxygen;
(g) treatment scheme planned or chosen: gas discharge to the atmosphere after ozone decomposition or recycling of the latter, in particular when oxygen is used.

Whatever system is adopted, certain elements of the equipments are common to the three types of installations; in particular:

(a) ozonated carrier gas collection system from the contact columns;
(b) gas conveying pipes;
(c) systems for placing the gas under a slight pressure so as to overcome head losses in the disposal system;
(d) vacuum-breaker system to be inserted between the columns and the suction system;
(e) air vent or recycling pipes after the decomposition system;
(f) the measurement or monitoring instruments, whose nature and incidence are approximately the same in the three processes, as will be seen further.

THERMAL DECOMPOSITION

The destruction of ozone by heating is based on the fact that this gas breaks down under the effect of heat (see Figure 1). A breakdown of more than 95 % requires temperatures of the order of 230–240°C, at which the destruction rate usually reaches 98 % for a reaction time of 1–2 seconds.

Concerning the three types of decomposition reported here, it must be pointed out that with the residual ozone concentrations we are dealing with, the equipment and power required do not depend on the concentration of the ozone to be disposed of, but only on the air flow through the disposal system and the initial conditions under which it is collected (pressure and temperature), the ozonated air's temperature being assimilable to the water's temperature in the contact columns.

Assuming a specific heat of 0.2454 kcal . kg^{-1} at constant pressure and a temperature of t°C for the moist air leaving the contact towers, the temperature T°C required to destroy the quasi-totality of the residual ozone contained in 1 m^3 of air will be reached when the following amount of heat is supplied:

$$Q \text{ kcal} = (T - t) \times 0.2454 \times 1.293$$
$$Q \text{ kcal} = 0.3173\,(T - t)$$

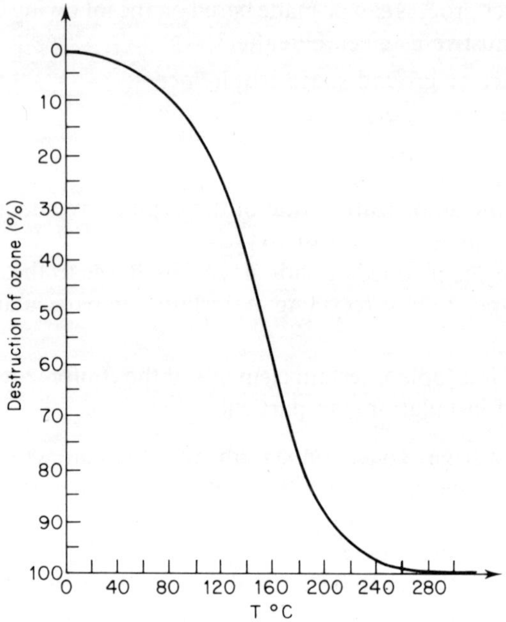

Figure 1

Where E is the electric power required per m^3 (STP) of air to be treated, and with 1 kWh = 861.29 kcal, the following equation can be written:

$$E \text{ kWh/m}^3 \text{ (STP)} = 368 \times 10^{-6} (T - t)$$

The rise in temperature of the gas to be treated is obtained using furnaces equipped with electric resistors, all elements being thoroughly insulated to prevent heat losses.

ONCE-THROUGH SYSTEM

In small furnaces, used for low flows, it is advantageous to design gas circulation such that heat transfer efficiency is increased and the time for temperature rise is reduced to a minimum. Countercurrent circulation systems permit these criteria to be met under good conditions. The same criteria are applicable for the furnaces used for higher rates, and the division of the furnace into three sections is interesting as far as energy is concerned:

I_1—preheating section; I_2—heating section; I_3—constant temperature section.

Such an arrangement permits the heating rate to be adapted to the gas flow by means of a simple temperature control system. Whatever type of furnace is used, the ozone-free carrier gas has a high temperature at furnace outlet. This implies, for average outlet, temperatures of about 250°C. In these once-through systems the discharge line sections must be oversized and refractory materials be used for the construction of these lines. Also, it must be pointed out that in view of the temperature, the carrier gas at furnace outlet conveys a fairly high amount of energy, proportional to the treated gas flow. This energy will be dissipated and lost in the environment.

All the above led ozone plant designers to develop systems for the recovery of part of the energy used for ozone destruction; the problem is the more acute as the gas flow to be treated and hence the electric energy required for destruction is higher.

THERMAL DECOMPOSITION WITH ENERGY RECOVERY

The configuration of such equipment is basically identical with that of the once-through thermic ozone decomposing system. Part of the energy content of the treated gas is recovered in a heat exchanger, designed for each particular case, and then used for the pre-heating of the gas to be treated.

A well-designed plant with high-yield heat exchangers allows savings of at least 50% on the amount of energy required for ozone destruction. The gas temperature at plant outlet is brought down to ranges at which traditional materials can be used and the oversizing of line sections is no longer a problem. In a well-designed plant, the gas temperature at the outlet, after recovery, ranges from 90 to 110°C depending on the operating conditions of the heat disposal system itself.

Cross-flow, gas–gas, shell-and-tube type exchangers or gas–gas, plate-type exchangers are the most used. Provision must be made for a system to place the gas to be treated under pressure to compensate for the pressure drop created by the insertion of the heat exchanger in the ozone disposal line.

The apparatus used for this purpose must, depending on the arrangement chosen, be capable of resisting the action of ozone in a moist medium or handling hot gases, and incidentally resisting accidental traces of ozone.

In view of the above, installing an energy recovery system is only advantageous, as far as capital costs and economy are concerned, when a certain ozone production is involved. The minimum hourly production rate from which this system is advantageous, is of about 4 kg ozone.

CATALYTIC DECOMPOSITION

The use of catalysts for excess ozone decomposition is fairly recent.

The first catalysts employed were not specifically designed for this operation. Their use involved a certain number of constraints and their lifetime while in service was rather short. To obtain correct efficiency along with a satisfactory service time, the air had to be preheated between 70 and 100°C as initial condition before passing through the bed, to prevent water condensing on the catalyst. Nitrogen oxides, chlorides and sulphides formed metal salts which resulted in passivating the catalyst, which was thus irreversibly deteriorated. The flowrates per volume per hour recommended for such catalysts ranged from 2,000 h^{-1} to 6,000 h^{-1}.

A new generation of catalysts was introduced on the market a few years ago: these were alumina and precious metals, specifically designed for ozone destruction. As far as yield is concerned, the heating temperature is reduced to 50°C, the flowrate per volume per hour is increased to the range of 4,000 to 10,000 h^{-1}. The pressure drop through the catalyst bed is reduced, and resistance to the influence of NO_x and Cl compounds is improved. Although these catalysts have a better yield, they remain sensitive to moisture and the presence of NO_x and Cl compounds along with moisture involve catalyst deactivation. This phenomenon is, however, reversible in this case and the catalysts can be regenerated by heating them at approx. 500°C for 6–7 h. It should be pointed out that sulphur compounds are poisonous to these types of catalysts.

Based on these products, certain manufacturers developed a third generation of much more efficient specific catalysts, trying to minimize or eliminate their failings. Nowadays, high-yield catalysts little sensitive to acid agents and having a much longer service life are available. Their main characteristics can be summarized as follows:

(a) higher flowrate per volume per hour: 10,000 h^{-1} as an average;
(b) lower heating temperature: 40°C;
(c) longer service time;
(d) very high resistance to sulphur and nitrogen oxide compounds;
(e) high resistance to small quantities of chlorine.

Although these catalysts' characteristics are improving, their prices have slightly decreased: thus, an ozone disposal system using catalysis has approximately the same capital cost as a heat system. Concerning operating expenses, the catalyst is more advantageous as the amount of energy required for preheating is much lower than the amount of energy consumed by a heat system.

In view of the small volume of catalyst needed for ozone destruction, and the rather low service temperature, ozone disposal systems using catalysis are compact and generate no nuisance due to heat.

FURTHER READING

W. Masschelein, Destruction de l'ozone, cout et securité dans les installations, European Forum on Ozone Applications to Water Treatment, Nice, 25–26 January (1979)

M. Pare and C. Coste, Cout d'exploitation de l'ozone et bilan energetique, European Forum on Ozone Applications to Water Treatment, Nice, 25–26 January (1979)

CHAPTER 47

A practical type of thermal residual ozone destructor

P. CHAPSAL
Trailigaz

A range of thermal residual ozone destructors has been constructed by Trailigaz to meet the needs of large installations treating water with ozone. These destructors are based on the principle of a total decomposition of ozone by raising the temperature of the gas to 350°C in electrically heated ovens. In this type of apparatus the essential economic factor is the expenditure on energy, which can represent a significant operating cost. This major concern should not diminish the other operating problems, which are: the accessibility to internal parts in contact with the gas for cleaning, the size of the equipment, the reliability of temperature control, durability and safety in operation. Generally, most of the heat produced by the electric resistors is recovered by means of a plate-type heat exchanger with a high degree of turbulence so that the air at the outlet of the decomposer, free from ozone, has a relatively moderate temperature of the order of 85°C. Moreover, all the external parts are provided with efficient heat insulation, some of which can be removed to give access to the interior of the oven.

Figure 1 shows such a destructor in diagrammatic form and, in particular, the routing of the gases through the heat exchanger and the electric pre-heater, as also the principle of the construction of this heat exchanger. The over-all efficiency of this system depends on the design of the equipment, which must have a high heat-exchange coefficient. Various manufacturers are at present offering this type of exchanger with a series of tubes, provided with plates having a section appropriate to promote a favourable degree of turbulence.

Progress has been made in regard to these systems because of the necessity of recovering the heat that was formally lost in various industrial processes using thermal fluids.

As a practical example, a Trailigaz destructor capable of treating 1350 Nm^3/h, corresponding to an installation with a production capacity of 15 to 27 kg O_3/h.

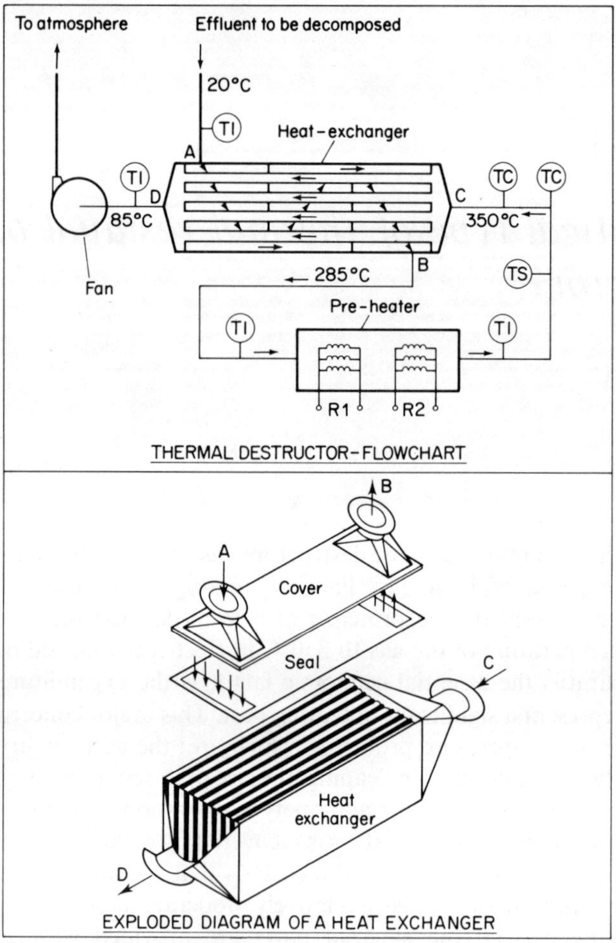

Figure 1. Destructors—heat exchangers

Overall dimensions are 3.2 × 1.8 × 2.3 m and weight about 2800 kg. Figure 2 shows this unit equipped with heat insulation, ready to be installed.

The metal parts used in this construction are of stainless steel. The overall loss of head in the circuit is 500 mm CE at a maximum flowrate of 1300 Nm^3/h and a theoretical contact time of 3 seconds. The electric power installed is 72 kW, which requires 20 minutes for starting up and reaching thermal equilibrium (for a 300°C internal oven temperature). The control device acts on and monitors the power absorbed by the heating resistors, in permanent operation, and when the gas flowrate varies according to the diagram of the ozone installation.

The average consumption of electric energy under normal operating conditions is 35 Wh/Nm^3 of gas.

These destructors are used in many ozonization installations, including the potable water treatment plants of the following cities: Helsinki, Marseille, Haamstede (Netherlands), Nice, Paris/Ivry, and Paris/Choisy.

Figure 2. Thermal residual ozone decomposers

CHAPTER 48

Removal of the residual ozone in the air after the application of ozone

M. HORST

Benckiser Wassertechnik GmbH Schriesheim

In all known applications where ozone is used for water treatment there is one main problem: the applied ozone cannot be dissolved totally in water; on the other hand—with respect to the mechanism of reactions—the consumption of ozone is not 100 %. The consumption of ozone in waterworks lies in the range of 91–99 % maximum, in water treatment of swimming pools in the range of 80–90 % and in wastewater treatment only in very special applications to nearly 99 %. The three most-used methods for removing ozone are:

(a) thermal destruction;
(b) destruction with activated carbon;
(c) catalytic destruction.

THERMAL DESTRUCTION

As ozone destruction is very sensitive to catalytic influences, the facts reported in the literature are rather conflicting. Therefore it is important to define the test conditions very exactly. In our test arrangement we sent $4\,m^3/h$ air containing ozone in a concentration of 2 to 5 g/m^3 through a stainless steel tube of 1 m length. The data found on heating up are shown in Table 1. To destroy the ozone completely air has to be heated up to 310°C.

DESTRUCTION OF OZONE WITH ACTIVATED CARBON

The consumption of activated carbon takes place by following main reaction of carbon burning

$$2\,O_3 + 3\,C \rightarrow 3\,CO_2$$

Table 1

Temperature of gas output (°C)	O_3 destruction (%)
50	0
100	10
150	21
200	34
250	50
300	99.9
310	100

That means 2.7 g ozone will react stoichiometrically with 1 g activated carbon.

Also, mechanical decomposition of the macro-structure of activated carbon and chemical decomposition of the macro-structure cause the formation of powdery activated carbon which tends to explode. Therefore activated carbon has to be changed at regular intervals. Moistening is an alternative method to avoid explosions.

CATALYTIC OZONE DESTRUCTION

A large number of catalysts for ozone destruction are known; for example some heavy metals, metal oxides, hydroxides and peroxides. In ozonization praxis, a catalyst made of aluminium oxide which contains palladium gives good results. This catalyst has to work inside a temperature range between 50 and 70°C for reasons of increasing the efficiency and avoiding inactivation which may be caused by condensated water. Under those working conditions this catalyst is working without any maintenance for a year.

The advantages and disadvantages of the three methods are outlined in Table 2.

COMPARISON OF RUNNING COSTS FOR TWO DIFFERENT EXAMPLES

First example

100 m^3 waste air/h

2 g O_3/m^3

(a) thermal ozone destruction without heat-exchange recovery:
running costs: about DM 15,000 (or 7500 US$).

(b) removal of ozone by activated carbon:
running costs: about DM 2000 (or 1000 US$).

(c) catalytic ozone destruction:
running costs: about DM 1760 (or 880 US$).

Table 2. Comparison of the advantages and disadvantages of the different methods for removal of ozone

Method	Advantages	Disadvantages
Thermal removal of ozone (once-through method)	(a) Complete ozone destruction independently of the O_3 concentration (b) No additional environmental toxicity involved (c) Operation without, or with little, maintenance	(a) Very hot waste gas (310°C) (b) High operation costs (c) High investigation costs
Removal by means of activated carbon	(a) Cool waste gas (b) Low investigation costs (c) Low running costs in case of low O_3 concentration	(a) Unpopular handling (b) Detonation danger (c) Additional environmental toxicity is possible (CO, CO_2) (d) High running costs in case of high O_3 concentration (e) Intensive service
Catalytic removal of ozone	(a) Long lifetime in adherence to the running conditions (operation temperature 50°C) (b) Operation without maintenance (c) Relatively low running costs in case of high O_3 concentration	(a) Sensibility in case of catalysts poisoning and humidity (b) Relatively high investment costs

For the computation of the operational cost of thermal destruction following parameters are considered:

$$C_{p(\text{air})} \approx 2.8 \times 10^{-4} \frac{\text{kWh}}{\text{kg} \times \text{degree}} \quad \text{and} \quad \rho_{(\text{air})} \approx 1.3 \frac{\text{kg}}{\text{m}^3}$$

the calculation delivers an energy consumption of 10.56 kWh to heat up 100 m^3 of ozone-containing air from 20°C to 310°C. Paying attention to the facts that the air also containing humidity and air heating is working with an efficiency smaller than 1, one should add 30% to the energy consumption. The total energy consumption is then 14 kWh. In the case of a hypothetical ozone concentration of 2 g/m^3 the resulting energy consumption per g ozone is 70 Wh. If the price for electrical energy is DM 0.12 per kWh one gets running costs of about DM 15,000 per year (or 7500 US$). As operation with activated carbon is not reliable and very different from case to case only empirical values can be given. Situations are known where the breakthrough happens when 10% of the theoretical efficiency has been reached. Calculations with stoichiometric reaction and neglecting effects based on mechanical and chemical destruction of the macro-structure calculation leads to a consumption of activated carbon of 0.37 g for removal of 1 g ozone. If the price for 1 kg of activated carbon (in Western Germany) is about DM 3 (= 1.5 US$) whence the running costs in this example are about DM 2000 (= 1000 US$) per year. The latter, compared with those of the thermal destruction, are quite satisfactory. However the maintenance is expensive.

Ozone destruction is a heat-producing reaction: the destruction of 1 mol ozone delivers 144 kg Joule. The 100 m^3 ozone-containing air in the mentioned example will be heated up by 4.4°C. Taking into account that the temperature of the off-air is 20°C, the reaction gas has to be heated up by other 35.6°C because the catalytic ozone destructor has an optimal operation temperature of 60°C. The energy consumption of 1.3 k Wh costs about DM 1360 (680 US$) per year. The costs of 20 kg of catalysts are about DM 2000 (1000 US$). With a hypothetical lifetime of the catalysts of about 5 years there are a further DM 400 per year. Total costs are then DM 1760 per year (or 880 US$ per year).

Second example

The choice of the best ozone destructor may strongly depend on the special application, running costs for a different example can be given as follows:

10 m^3 of ozone containing air have to be treated per hour. The concentration of ozone in this case is 20 g O_3/m^3.

(a) thermal ozone removal:
running costs: about DM 1040 (= 520 US$).

(b) removal of ozone by activated carbon:
running costs: about DM 2000 (= 1000 US$).

(c) catalytic ozone destruction:
running costs: about DM 400 (=200 US$).

With the same data as in the first example there is an energy consumption of about 7 Wh for the removal of 1 g ozone. The costs per year continuous operation are DM 1400 (700 US$).

Since the consumption of activated carbon depends on the absolute amount of ozone which has to be destroyed the running costs are the same as in the first example, that means DM 2000 per year (or ≃1000 US$).

The catalytic destructor equipment being well isolated, the heat produced during the reaction of ozone destruction is sufficient to keep the catalyst at its optimal operation temperature. There is no need of additional energy consumption. Running costs are reduced to the costs of exchanging the catalysts every 5 years, that means DM 500 per year (≃250 US$).

CHAPTER 49

Methods and operating costs of ozone destruction in off-gas

K. ORGLER

Trailigaz

The ozonized air leaving the contact columns contains a small amount of ozone which must be removed before discharging the effluent into the atmosphere.

Ozonization plants today basically make use of three methods for decomposing the ozone in off-gas. These are:

(a) the catalytic method;
(b) the thermal method;
(c) the adsorption and decomposition on activated carbon.

DECOMPOSITION OF OZONE BY CATALYSIS

The catalysts, based on metal oxides, which are extruded, are available as extruded granules in the form of 'threads', measuring 4 × 4 mm (Degussa) or 4 × 6 mm (Kali-Chimie). They are designed for the quantitative decomposition of ozone, even in cases of very high concentration exceeding 30 g O_3/m^3 of air.

Water weakens their mechanical resistance and catalytic activity. The catalysts must work at a temperature which prevents their humidification by condensation, which may appear as a result of the decreases beyond the dew point of the ozonized air saturated with water allowed to reach the contact mass.

The design of the catalytic ozone decomposing system is based on the spatial velocity (VS h^{-1}), expressing the ratio of the flowrate of the ozonized air to be treated, Q (m^3/h), to the volume of the catalyst V (m^3) required.

To prevent water condensing on the catalyst, it is recommended that the temperature and the flowrate of the ozonized air be regulated and monitored. If wetted accidentally, the catalyst can be reactivated by careful heating up to 120°C, but the temperature must never exceed 130°C. Nitrogen oxides, chlorine and its derivatives and sulphides have a chemical action on the catalyst. This is expressed by passivation due to the formation of metal salts, the resulting poisoning being permanent.

Metal catalysts are always more active than oxides or combinations. Good advantage is taken of this in the catalytic method of decomposing ozone, and by the use of platinum or palladium in finely divided form on the supports, giving good mechanical strength. The operating temperature is 30 to 70°C, thus lower than that recommended for the metal oxides.

DECOMPOSITION OF OZONE BY HEAT

The decomposition of ozone by heat in gaseous phase in a dynamic system, up to a residual value of about 0.1 ppm, is attained at a temperature of 350°C for a contact duration time of 3 seconds (see Figure 1).

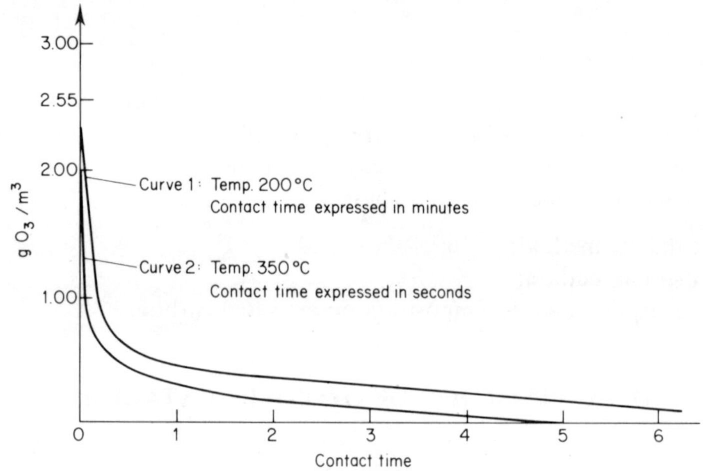

Figure 1. Thermal decomposition of ozone

The heat can be transferred in several ways to an off-gas stream, e.g. through flowing over electric resistances with insulated wires or through combustion of oil or natural gas inside the heat exchanger with direct heating. In order to recover most of the heat used, a heat exchanger (Figure 2) must be included in the cold and hot line loop, thus solving the problem raised by recent regulations on energy saving.

REMOVAL OF OZONE BY ACTIVATED CARBON

The removal of ozone by means of activated carbon is based on adsorption, accompanied or not by reaction and possibly a catalytic process.

Use is made of dry or moist filters, containing carbon in granulated or thread form, measuring between 1 and 6 mm.

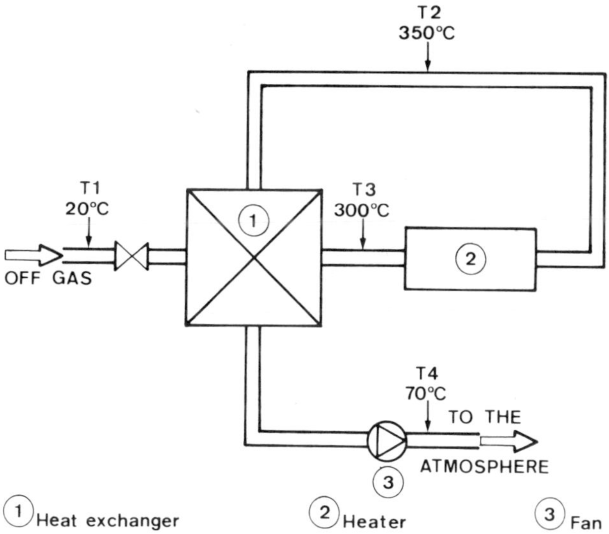

Figure 2. Operation of an installation for decomposing ozone by means of heat

COST OF DECOMPOSING RESIDUAL OZONE IN OFF-GAS

It is not possible within the scope of the present contribution to discuss thoroughly the cost of the various processes for decomposing ozone presently in use, and all the more so since they depend on both general and local factors.

Table 1 is a comparison (French francs and US dollars) of a treatment of 120 m^3/h of ozonized air escaping in off-gas.

Table 1

Item	Operating cost per year	
	FF	US$
Oxide catalyst	4,100	700
Metal catalyst	2,840	460
Thermal, electric decomposition		
without heat-exchanger	18,790	3,130
with heat-exchanger	3,800	630
Thermal decomposition with fuel oil		
without heat-exchanger	15,600	5,200
with heat-exchanger	3,250	540
Thermal decomposition with natural gas		
without heat-exchanger	8,000	1,350
with heat-exchanger	2,000	670
Decomposition by activated carbon	7,000 to 8,000	1,300

CHAPTER 50

The Tailfer plant and new developments in ozone destruction techniques

W. J. MASSCHELEIN

Compagnie Intercommunale Bruxelloise des Eaux

Previous experience of the CIBE (Brussels Intercommunal Waterboard) in the off-gas treatment processes was reported, e.g. at the I0A-Symposium at Nice.

The Tailfer plant of the CIBE in its present evolution is designed to operate with a maximum nominal gas-flow of 1220 Nm^3/h which is injected in four treatment lines each subdivided into two parallel files. The ozone production ranges 24 kg/h (maximum) when using air and can be uprated to 42 kg/h (maximum) when using an oxygen-enriched process gas. The ozone losses at the chimney are below 1 ‰, i.e. 0.2–0.3 g/Nm^3, and can occasionally attain 0.5 g/Nm^3 for short periods. The ozone production lines are subdivided into six files of 220 Nm^3 each.

Presently, the installed technique is a thermal heat exchanger warming-up the off-gas at a temperature of 300°C in the active zone of the furnace constructed in stainless steel (AISI.316). The exchangers installed are able to treat 400 Nm^3/h and can be uprated to attain 600 Nm^3/h as given in the general scheme in Figure 1.

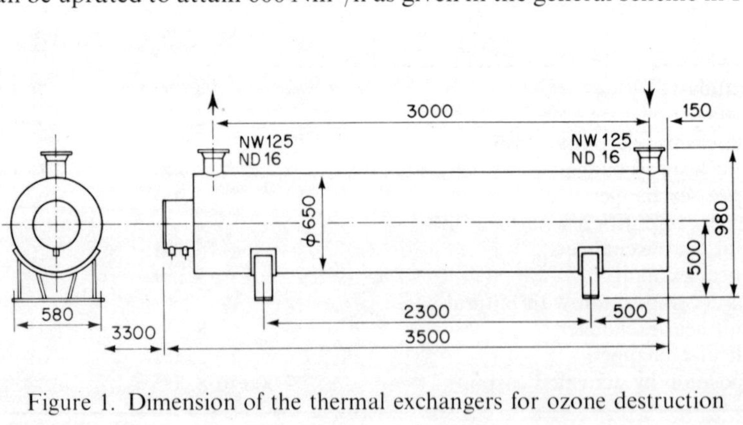

Figure 1. Dimension of the thermal exchangers for ozone destruction

At the exit, the gas temperature is 90–100°C at nominal flow and the head-loss of the equipment can attain 0.1 bar. At nominal capacity, i.e. 400 Nm^3/h, the energy consumption ranges 17 Wh/m^3. The destruction yields attain $\geqslant 99\,\%$. The method operates adequately but remains very expensive especially at the present increase-rate of energy costs. Thus further research may be required to increase the yield at decreasing costs of the ozone destruction in the off-gases. The unit is shown in Figure 2. Table 1 gives the details of the costs, on a comparative basis of 220 Nm^3/h as is operational in the Tailfer plant.

Figure 2. Thermal destruction unit installed in Tailfer

The following alternative methods are worth mentioning, as showing promise.

The preconcentration of ozone contained in the off-gas by adsorption on solid supports (e.g. appropriate synthetic zeolithes) is under investigation so as to establish the endurance of the sorbing agents as well as the safeness regarding explosivity. In some cases, the results remain favourable provided the temparature during the desorption–regeneration attains at least 240°C.

Another method experimented recently on a pilot scale in the Tailfer plant is the photolysis of ozone induced by UV (254 nm), irradiation.

The photochemical decomposition of ozone in the gas phase is possible through a wide range of frequencies from the UV to the IR. In the UV zone ^1D oxygen radicals can be produced capable of being deactivated by a bi-particle contact, e.g. with another O radical, ^1D or ^3P, or through collision with the wall.

Table 1. Comparative costs to treat a unit flow of 220 Nm^3/h (prices in BF (Belgian francs) 1980 converted 40 BF $\simeq$ 1 US$)

	Installation costs		Working costs
	BF	US$	(kWh)
Electrical resistance	620,000	15,500	27–34
Heat exchanger	1,475,000	36,900	17–19
Combustion	740,000	18,500	6†(xx)
Activated carbon	620,000	15,500	3
Absorption + thermal destruction (or catalysis)	740,000	18,500	9
Chemical scrubbing*(x)	1,250,000	31,300	2–4
U.V. photolysis	500,000	12,500	12

* Still under investigation
† The cost of fuel is computed in equivalent kWh (1980)

According to the literature, the quantum yield in a dry process gas would hardly attain 2 but the water partial-pressure is critical. In these circumstances $OH^{\cdot}$ radicals capable of reacting with the $O(^1D)$ radicals are formed also, and by a chain reaction mechanism the quantum yield can attain values up to 130 % (see Figure 3).

In the latter, the fundamental primary reactions initiating the UV decomposition of ozone are accepted as:

$$O_3(^1A) + h\nu \rightarrow O_2\left(\overset{1}{\sum} g\right) + O(^1D)$$

$$O_3(^1A) + h\nu \rightarrow O_2\left(\overset{1}{\sum} g^+\right) + O(^1D)$$

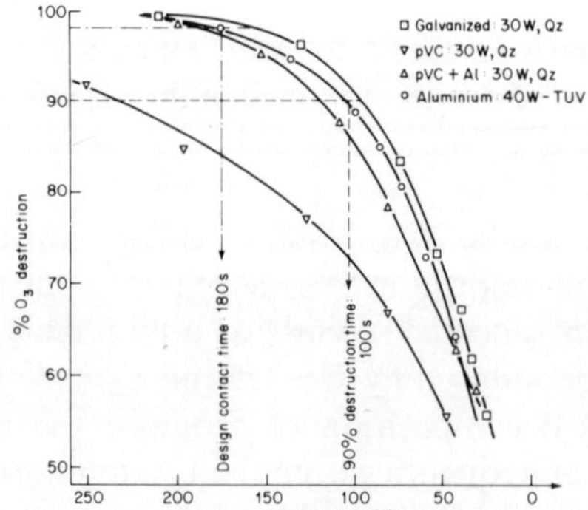

Figure 3. Yield curves of UV-decomposition of ozone

Guidelines for design of the UV method are actually set as follows:

(a) The use of square contact chambers (or multiple-tube chambers) by placing the tubes in a square arrangement. The maximum irradiation depth is set at about 10 cm with reflection chambers.
(b) The inside walls are best coated with an UV-reflecting material, e.g. aluminium (the resistance of this material is still being tested).
(c) Low çost optical-glass UV lamps based on the 254 nm low-pressure gaseous mercury emission through fluorescence are satisfactory.
(d) The nominal transit time or gas exposure time should be ± 180 s for a 98–99 % destruction and ± 100 s for 90 %.

In order to treat 220 Nm^3/h on a continuous basis, the annual removal of the tubes may implicate an additional yearly cost of 360,000 BF (1980) (9000 US$). Thus for large plants the method appears expensive but can be suitable for small plants owing to the simplicity of the technique. As for chemical scrubbing techniques, they have not yet been fully investigated.

FURTHER READING

W. J. Masschelein, G. Fransolet and B. Avalosse, *Symposium IOA—Europe—Nice*, IOA-Europe, Paris (1979)
W. J. Masschelein. In, R. G. Rice and A. Netzer, to be published by Ann Arbor Michigan (1981)
W. J. Masschelein, *IOA-Congress, Berlin*, IOA, Paris (1981)

CHAPTER 51

Bacterial stabilization of ozonized water

H. LAMBLIN

Compagnie Générale des Eaux

Ozone is a powerful bactericidal and viricidal agent. However, unlike residual chlorine, residual ozone in water disappears quickly—especially when temperature and pH are increased: as for instance at pH 7.6 and a temperature of 20°C, residual ozone of 0.4 mg/l disappears within an hour. This disappearance assures that no taste or odour of ozone will be detectable. Consequently residual ozone cannot be relied upon to assure the bacterial stability of treated water in transit in the distribution system. Other measures must be taken during treatment to improve the quality of ozonized water for the final sterilization before distribution. These measures will depend on the nature of the raw water.

THE CASE OF WATER CONTAINING AMMONIACAL NITROGEN

Some particularly resistant non-pathogenic germs seem to escape ozonization treatment applied in the usual conditions. This is especially true for nitric and nitrous germs. Where ammoniacal nitrogen exists in a distribution network, these germs may begin to develop considerably and then accumulate on the conduit walls, sporadically breaking off and causing red water at the consumer's tap. Ozone has practically no effect on ammoniacal nitrogen, so the treatment plant must include a stage for its elimination by biological or chemical action. If the raw water's ammonia content is high and variable over short periods of time, it is advised to include a final post-chlorination after ozonization to correct possible imperfections of the ammonia elimination process before distribution.

THE CASE OF WATER LOADED WITH ORGANIC MATTER

When post-chlorination takes place after ozonization, the total chlorine residual is considerably less stable than it is without ozonization. This is true in first instance for water loaded with organic matter. It seems that ozonization partially transforms non-assimilative organic matter into assimilative by-products that can then serve as nutritive matter for micro-organisms.

A well-adapted solution to this problem consists of completing ozonization treatment with filtration on granular activated carbon. Organic matter rendered biodegradable by ozone is treated biologically in a carbon filter. This combination of ozone and activated carbon then can increase the elimination of organic matter and provide a treated water of higher quality, and of greater bacterial stability.

In the absence of activated carbon filtration, the chlorine demand must be met after ozonization by adding sufficient chlorine to provide an active residual.

PROTECTION OF A DISTRIBUTION SYSTEM AGAINST BACTERIA

In the case of a short network where water temperatures remain low, the bacterial quality of water can be maintained in transit with a final ozonization, without either post-treatment or sterilization with an active chlorine reagent.

But in the case of an extended and interconnected network, or where water temperatures can reach high levels, additional treatment for disinfection is necessary after ozonization to avoid bacterial growth in the network.

The choice of chlorine reagent and its point of application both depend on the necessary dosage to maintain an active residual concentration and the characteristics of the distribution system.

Chlorine dioxide has the advantage of being relatively stable and of causing less odour than chlorine. However, it does not act on ammonia. It can be used only at very low doses so as to avoid too high concentrations of chlorite. Moreover, seeing that ozone decomposes chlorine dioxide, the injections must take place after the disappearance of residual ozone. For the above-mentioned reasons it is not advised to surpass a dosing of 1 g/m^3 ClO_2 in post-treatment and ± 0.5 g/m^3 can be detected by odour.

Chlorinated water or hypochlorite can also be used for post-chlorination. When large doses are used, the drawback of water taste can be reduced by dividing the sole injection of the chlorine reagent into chlorine and chlorine dioxide injections.

The fixing of the doses of a chlorination post-treatment for the protection of the network is a delicate problem. The method most commonly used until now obeyed the following rule: the presence of a chlorine residual must be detectable at all points of the distribution system. For very extended networks the application of this rule can lead to very high dosages, exceeding 1 mg/l in some cases. The disadvantage of this method is that it entails an increase in unpleasant tastes and odours, especially for consumers near the treatment plant.

This situation is even more paradoxical for water plant operators who use ozone to improve the organoleptic qualities of treated water.

So more sophisticated means than residual chlorine control in the distribution system must be implemented. A possible solution is to study the evolution of bacteriological characteristics of treated water using, as indicators, the total germs cultivating at 37°C in 24 hours, at at 20°C in 72 hours.

To control the evolution of treated water and to determine whether treatment was well applied, samples of treated water are conserved in flasks and undergo enumeration at regular intervals. Treatment irregularities can thus be promptly detected and corrected.

These enumerations are carried out on samples taken in various locations representative of the distribution system. This control is used to increase or decrease post-treatment chlorination in view of the evolutionary tendency of the plate-count germs. The use of plate-count germs as indicators of the bacterial quality of distributed water is more elaborated than that of the residual chlorine content, as the usual value is not zero. Post-treatment chlorination rates can be reduced substantially, avoiding the degradation of the organoleptic qualities of ozonized water.

FURTHER READING

W. J. Masschelein, Maintien de la potabilité des eaux dans les réseaux de distribution, *Bull. Inf. Anseau*, 121 (1975)

H. Lamblin, Protection d'un réseau d'eau ozonée vis-à-vis des germes, *L'eau et l'Industrie*, no. 17, pp. 35–44 (June–July 1977)

J. L. Biros and H. Lamblin, Reduction of residual chlorine by intensive control of bacterial population in supplied water: a study on the Paris suburbs distribution system. In: *Water Chlorination—Environmental Impact and Health Effects*, vol. 3, R. L. Jolley (ed.) Proc. 3rd conference on environmental impact of water chlorination, Colorado Springs, 28 October–2 November 1979, Ann. Arbor Science Publication, Ann Arbor (1981).

CHAPTER 52

Some aspects of bacterial aftergrowth in ozonized waters

G. FRANSOLET

Compagnie Intercommunale Bruxelloise des Eaux

In practice ozonization, as performed in water treatment, not only releases carbon dioxide and oxidized nitrogen compounds but mostly partially oxidized organic products such as carboxylic acids. These can result from the oxidation of the phenolic nucleus and can influence bacterial growth in two different ways: firstly by the elimination of bacteriostatic or bactericidal compounds, and secondly by the formation of acids metabolizable in the tricarboxylic cycle. This effect has been evidenced with several strains including NCIB typical *Pseudomonas*.

In all instances a correctly ozonized water should always be free of bacteria indicating a fecal pollution such as coliformes, *E. coli*, and enterococci.

A typical experiment to check the bacterial evolution in a water is to conserve the water in (closed) bottles and proceed to counting at regular intervals between 3–7 and 15 days. As a general rule, in the case of a well-ozonized water, although it may contain about 10 germs/10l, no significant aftergrowth is observed; which is not the case of groundwater which, although disinfected with 0.2 g Cl_2/m^3, gives rise to a systematic bacterial development. The probable conclusion of this investigation (still being carried out) is that an ozonized water contains less or no germs capable of aftergrowth as compared with a groundwater. Hence, mixing of both types of water can to some extent be considered as a seeding of the ozonized water. Furthermore, it is not possible to guarantee the complete sterility of all installations which store and transport the water. Vent pipes are sources of possible contamination, so are the access structures.

Among the bacteria implicated in aftergrowth, species of the germs Pseudomonadaceae, Nitrogen bacteria, micrococci and Coryneiform bacteria are to be found. Most of the yet isolated bacteria are of telluric origin which explains their ability of adaptation to grow on poor mineral media.

In spite of the fact that ozonized waters contain assimilable organic compounds the latter are present at only low concentrations. In this precise case the multiplication curves of the growth rate as a function of the substrate concentration are represented by a hyperbolic function as is given by the Monod's equation:

$$r = r_{max}\frac{C}{C_1 + C}$$

in which r and r_{max} are respectively the growth rate and the maximal growth rate; C the concentration of the limiting nutrient factor and C_1 the concentration at which the growth rate is equal to half of r_{max} (see Figure 1).

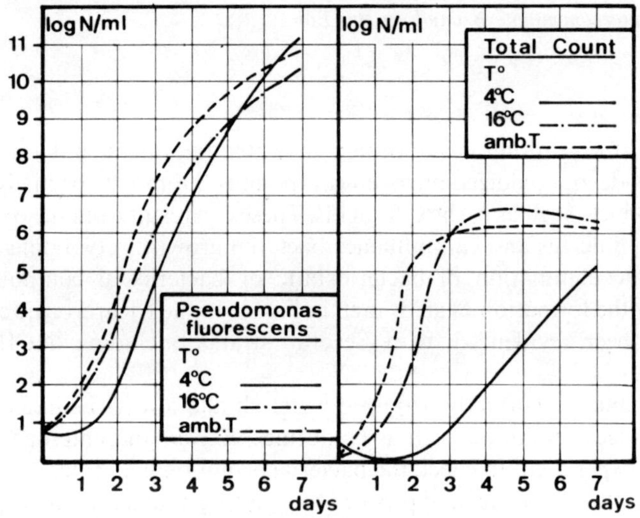

Figure 1. Typical aftergrowth curve

In most practical investigations the final stationary concentrations in germs reaches 10^6–10^7/ml. For total counts while the counts of ubiquitous *Pseudomonas* can reach up to 10^9–10^{11}/ml. These values are more or less independent of the type of water but are reached after variable delays and generally faster (e.g. within 3–4 days, in ozonized waters).

In practical systems, the limiting substrate varies and different products can be used by an organism on a sequential basis owing to inductive and depressing (or inhibiting) mechanisms. Also the competition in mixed bacterial populations cannot be overlooked. Therefore the consumption rate of a given substrate not only depends on the number and type of substrate but also on the interactions between the several species. Phenomena such as co-oxidation intervene.

The behaviour of mixed populations in the presence of various substrates depends on the similarities in affinities for specific (limiting) substrates. The principle of competitive elimination applies in the case in which two organisms exhibit similar affinities for the same limiting substrate(s) in their nutritional source. Also the diversity in available substrate is a predominant factor determining the heterogeneity of developing bacterial populations. The more the substrate is limiting the more the competition between two organisms is important.

Thus the bacterial aftergrowth, more particularly in ozonized waters, is far more complex than considered at first sight.

The temperature-effect may be an important factor especially through the relation between the temperature and the growth rate which corresponds to the Arrhenius' law as is the case for other biochemical processes like enzymic activity. In the neighbourhood of the optimal developing temperature an increase of approximately 10°C doubles the growth rate. Most bacteria isolated in experimental aftergrowth are of the mesophilic type and have an optimal developing temperature between 20 and 35°C. Consequently, for some strains at 10°C the optimal value is observed while for others, having an optimal temperature of 20°C, the growth rate can still represent 60 % and more of the optimal value when developing at 10°C. Further investigation on the phenomena of bacterial aftergrowth, including the identification of the organisms, is still required.

FURTHER READING

W. J. Masschelein, Le maintien de la potabilité des eaux dans les réseaux de distribution, *Bull. Inf. Anseau*, 121 (1975)

G. Fransolet, Aspect de la reviviscence bactérienne dans les eaux ozonées, *Bull. Inf. Anseau*, 157 (1979)

J. W. Foster, Hydrocarbons as substrates for microorganisms: Antonie van Leeuwenhoek, *J. Microbiol. Serol.*, **28**, 241–74 (1962)

J. Perry, Microbiol. Cooxidations Involving Hydrocarbons, *Microbiol. Rev.*, **43**, 59 (1979)

W. J. Masschelein, G. Fransolet and E. Debacker, Signification de la palingénésie microbienne et le moyen de l'entraver dans les eaux de distribution, *Eau du Québec*, **13**, 289 (1980); **14**, 41 (1981)

6. OZONIZATION COST

CHAPTER 53

Global cost evaluations

M. Paré

Degrémont

The costs associated with ozone production and use in water treatment can be schematized by the following phases:

(a) feeding with gas: air or oxygen;
(b) air drying or oxygen recycling, depending on type of feeding;
(c) ozone production;
(d) contact between ozonated air and water to be treated;
(e) optional disposal or recovery of excess ozone.

The estimation of the total ozonization cost must take into account capital costs and operating costs, the latter including energy expenses as well as maintenance and running personnel charges. Seeing that these different parameters vary in large proportions, depending on the economic conditions, the countries and the local traditions, it is only possible to indicate general guidelines. These correspond to a general procedure which is to be followed to assess the incidence of the costs of the various elements of an ozone production line, considering that the optimization of a plant amounts to defining an optimum ozone concentration in the carrier gas.

The working concentration, and hence the gas flow through the plant, have opposite effects depending on whether the gas conditioning and contact portion or the ozone generation portion is considered. Capital and operating costs for gas conditioning increase with the amount of gas while simultaneously ozone generators give better performances. The determination of a compromise must, therefore, be the aim of all economic study in terms of generation and use of ozone. Such a study was already carried out, more thoroughly, during the 'Internationales Symposium Ozon und Wasser' held in Berlin in 1977. As operating costs depend heavily on energy expenses, the energy-consuming units shall be taken into consideration directly as a cost parameter.

In the present case we have considered units using air, which is the most frequent and able to produce between 1 and 30 kg ozone per hour.

AIR DRYING

Two methods are available: (a) high-pressure drying using compressors working at a relative pressure of from 5 to 7 bars, and (b) low-pressure drying compressors using blowers working at a relative pressure less than 0.8 bar. In the case of low pressure drying, and at rather large plants, the dryer by adsorption is preceded by a cooling unit.

Although, under certain conditions, high-pressure drying is competitive from an investment viewpoint, it is inadvisable to adopt this method for large plants in view of its high energy consumption. Cumulative energies required for an air-conditioning system using low pressure, including air blowing, generation and the cooling units amount to approximately 0.06 kWh/Nm^3 of air useful for ozone production. When high pressure is applied, the energy consumed is about 0.1 kWh/Nm^3 which is almost twice the preceding value.

In view of the above, high-pressure drying is only employed at small or medium-sized plants since it is easy to use, and its low space requirements allow the construction of packaged, monobloc units.

OZONE PRODUCTION

As already mentioned, ozone generators' efficiency is higher when the gas flow through the plant increases, i.e. when the working concentration decreases. This amounts to saying that for a given production, the size of units and the energy they will consume will decrease with concentration and thus, other things being equal, the price for the production units, including the power supply and indispensable monitoring instruments, may, according to plant size, be 10–20 % higher when, for instance, the concentration is increased from 1 to 2 %. The cost of energy consumed, still depending on plant size, will be increased by 20 to 15 %.

CONTACT WITH THE WATER TO BE TREATED

Depending on the mode of contact between ozonated gas and water, the unit may or not consume energy. When porous diffusers are used, the required energy specifically for the injection has already been accounted for in the compression unit, whereas mechanical contact provided by turbines implies extra energy cost. Anyway, the capital cost and energy cost for the contact unit vary in function of the gas flowrate, i.e. in inverse ratio to concentration.

DISPOSAL OF EXCESS OZONE

Several means of excess ozone disposal are available, which are the subject of other contributions in this book. The most frequently employed are disposal by heat with or without partial recovery of energy, and catalysis with pre-heating. In all cases, capital and operating costs vary as a function of the gas flow to be treated.

Recent advances in the field of catalysis make this procss competitive from the viewpoint of investment and advantageous as far as the operating costs are concerned. Thus, the capital cost for a disposal system using catalysis will be of the same order of magnitude as a plain heat system, and about 50 % lower than a heat system with energy recovery. Furthermore, its operating cost will be 60 % lower than in the first system and 40 % lower than in the second one.

RECOVERY OF EXCESS OZONE

One may be tempted to re-use excess ozone collected at the vent of contact towers for treatment. This requires the installation of moist ozone-resistant compressors or turbines, but does not exclude the use of a disposal system before discharging the off-gas into the atmosphere as the residual concentration is still far beyond the generally admitted limits.

According to the plant size and the amortization of equipment, the total cost for such systems accounts for 10–12 % of the over-all cost of the other units considered in the study.

The use of such a system could only be justified from the sole economical standpoint if the savings allowed on ozone production were of the same order of magnitude, but this cannot be the case in view of the residual concentrations obtained and the dissolving and corresponding efficiencies.

OPTIMIZATION OF OZONE CONCENTRATION IN THE GAS

Considering the above data, a choice can be made of the best adapted system for the various units mentioned in the study.

As an example, Figure 1 shows the relative values of the various costs for a given concentration and under the following amortization conditions:

(a) linear amortization over 15 years for rotating engines;
(b) linear amortization over 20 years for static equipment.

The same study has also been conducted for several other concentrations: curves such as those in Figure 2 have been obtained, which show a minimum cost for a given concentration.

The respective parts occupied by the various partial costs in the overall cost can be clearly deduced in Figure 3. The result would, of course, be different if it was decided to amortize capital costs before operating costs, or if energy costs varied significantly.

Another important parameter for the optimum working concentration is the configuration adopted for the production line. The example under study dealt with the simplest line in which a reasonable minimum of units ensures a given

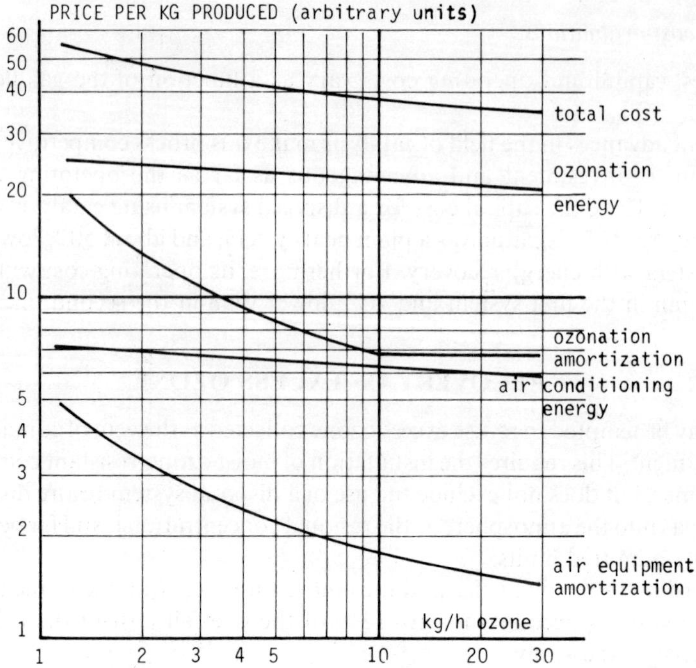

Figure 1. Relative values of the various production costs

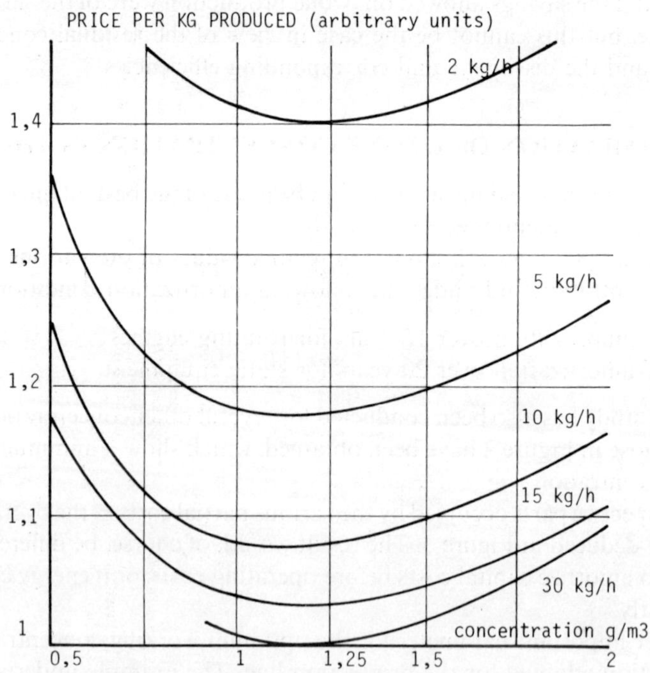

Figure 2. Optimum working concentrations

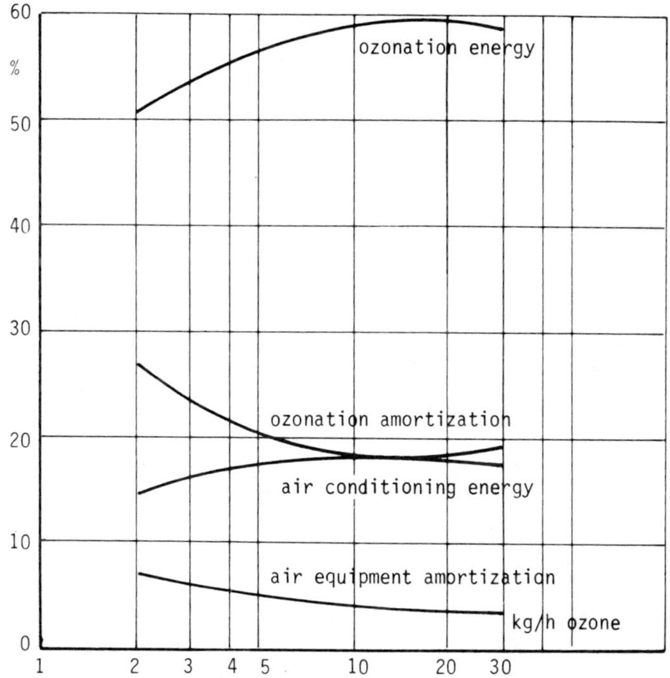

Figure 3. Relative values of the various production costs (the total of the four costs equals 100 %)

production. In other words, each equipment is supposed to be operated permanently. For another configuration, for instance including one or several standby ozonators, the optimum concentration would be less than that shown here so as to increase yields and hence reduce the cost of the ozone production unit as compared with the other units.

FURTHER READING

M. Paré, Cout d'exploitation de l'ozone et bilan energetique, *Internationales Symposium Ozon und Wasser, Berlin* (1977), O. H. Hess, Berlin, 1977.

CHAPTER 54

Cost–benefit analysis of the use of an oxygen-enriched process gas

W. J. MASSCHELEIN

Compagnie Intercommunale Bruxelloise des Eaux

Ozonization plants for drinking water treatment by the post-ozonization technique enable us to determine the global amortization costs (building + equipment to produce, inject and safely operate the plant): for instance 11 Mio BF (Belgian francs) to produce 1 kg O_3/h in a plant of a nominal total capacity of 5.6–6 kg O_3/h while a nominal capacity of 24 kg O_3/h using air implicates an investment of 5.5 Mio BF for 1 kg O_3/h capacity (all BF in 1980 rates, which at present currency is 40 BF $\simeq$ 1 US$).

The analysis of these two important plants suggests a relationship of the type

$$\left(\frac{C_1}{C_2}\right)^2 \sim \left(\frac{P_2}{P_1}\right)$$

In which P is the nominal production capacity of a plant using air and C the investment cost per hourly production capacity.

The global costs as well as the specific operation costs of the direct use of oxygen (i.e. without recycling) in the water ozonization are reported in Table 1.

Over-all design parameters for a cost-based design of the use of oxygen-enriched process gas have been reported before and can be summarized as follows:

(a) The permanent production capacity of an ozonator as 80 % of its maximal (nominal) capacity must be taken into consideration.

(b) This production capacity (80 % of nominal) must be capable to be uprated so as to meet maximal needs e.g. 200 % to $\simeq$150 % (average) compared to the minimal demand of 100 %.

(c) The maximum capacity able to meet maximal needs can be expressed as 80/100 nominal production capacity × 1.5 (1.75) which is the possible increase in production capacity obtainable by using a process gas enriched

Table 1. Operational costs for ozonization over 1978–1980 (BF/m^3) (40 BF ≃ 1 US$ 1981)

Year	Ozone production[1]	Contacting[1]	Oxygen[2]	Control and operation[3]
1978	0.031	0.02	0.027	0.075
1979	0.035	0.022	0.031	0.08
1980	0.044	0.027	0.016	0.075
Year	Maintenance[4]	Off-gas treatment[1]	Various[5]	Total
1978	0.047	0.01	~0.02	0.230
1979	0.048	0.01	~0.02	0.246
1980	0.046	0.02	~0.02	0.248

[1] 1 kWh (industrial coproducing tariff costs
in BF (1978): 1.11
in BF (1979): 1.16
in BF (1980): 1.34
For non-coproducers this can be multiplied by 225 %.
[2] Prices per Nm^3 O_2 in BF are in 1978 and 1979: 3.5; in 1980: 4.4.
[3] Also includes 10 % of the total laboratory costs related to the treatment.
[4] Twice a year + one permanent worker (assimilated) + spare tubes.
[5] Includes cleaning, venting, heating, etc.

with oxygen. Thus a nominal production capacity of 1.2 to 1.4 times the minimal needs becomes sufficient to meet a maximal ozone demand of 200 % of the minimum demand. This is equal to a scale economy of at least 1.3 to 2 or 35 % of the installed production capacity (which can reach 50 % reduction in some cases).

A typical example of seasonal fluctuation of the ozone consumption in the post-treatment of river water is given in Figure 1.

Table 2 shows the ratio of capital investments at the scale of the Tailfer plant in 1976. Reported on a yearly basis of 75 % of the nominal capacity of the plant, this is equal to a difference of 0.11 BF/m^3 or 3 to 4 times the operational cost of oxygen (see Table 1) as used according to the variable ozone demands of the river Meuse water.

Therefore, besides its operational advantages, the direct use of an oxygen-enriched process gas is confirmed to be economically feasible and to meet variable ozone demands, peak water consumption periods or other necessary variable operation rates of the plants.

Table 2. Ratio of capital investments at the scale of the Tailfer plant in 1976

Nominal production capacity	: 90 MF
Uprating oxygen by 75 % of nominal	: 10 MF
Economy of investments (0.75 × 90 − 10)	: 57,5 MF
Reactualization (1980)	: 84 MF
or an annuity of	: 8 MF

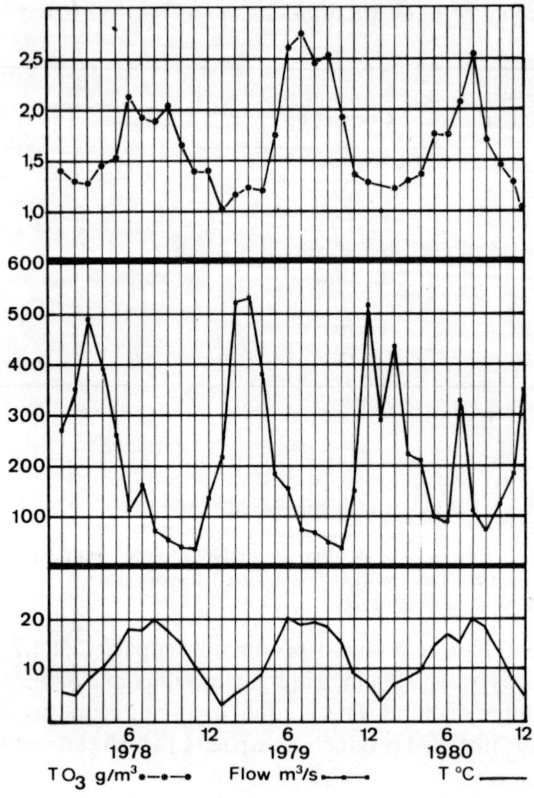

Figure 1. Variation in ozone demand of treated water (River Meuse)

FURTHER READING

W. J. Masschelein, *T.S.M.-L'eau,* **71,** 385 (1976)
W. J. Masschelein, *T.S.M.-L'eau,* **72,** 176 (1977)
W. J. Masschelein, *Congress IOA Berlin* (*1981*), O. H. Hess Verlag, Berlin, 1981, p. 434

7. DEVELOPMENTS IN OZONE USE

CHAPTER 55

Utilization of ozone in swimming pools

J.-P. LEGERON

Trailigaz

The increase in the number and size of swimming pools has resulted in the utilization of an ever-increasing quantity of water in the pools. Due to the increasing cost of the water, it has been necessary to restrict the supply: in other words, recirculation of pond water has been introduced for economical grounds and the recycled water is treated in order to eliminate pollution and hence provide an excellent quality of water in the pools.

The impurities brought in by swimmers and atmosphere can be classified into two categories: either (a) organic and mineral matters or (b) microorganisms.

Because of its different properties, ozone can play an important role in the treatment of recycled waters as it serves a dual purpose: oxidation and disinfection.

Figure 1 shows the different possibilities of ozonization as applied in the treatment of recirculated water. According to the place where ozone is injected, the process is called preozonization or post-ozonization respective to the filtration stage.

When introduced before the filters, ozone acts mainly as an oxidizing agent and it creates a more or less large quantity of flocs which are then separated. If a flocculating agent is used, the micellization–demicellization process takes place: ozone contributes to save reagents and helps to ensure a good elimination of organic matters. The latter can also be improved by biological filtration. In this case, ozonization has double objective: water oxygenation on the one hand, and improvement of the biodegradability of organics on the other. In the filter, in addition to the traditional metabolic schemes of organic matter nitrification also takes place, whereby the problems associated with ammonia are partially solved too.

When ozone is introduced after filtration, it can contribute to water disinfection. In this case, a residual of free dissolved ozone in water of 0.4 g/m^3 is maintained during a period of 4 minutes. In actual practice, this takes place in detention chambers composed of three sections, pressurized or not: in the first

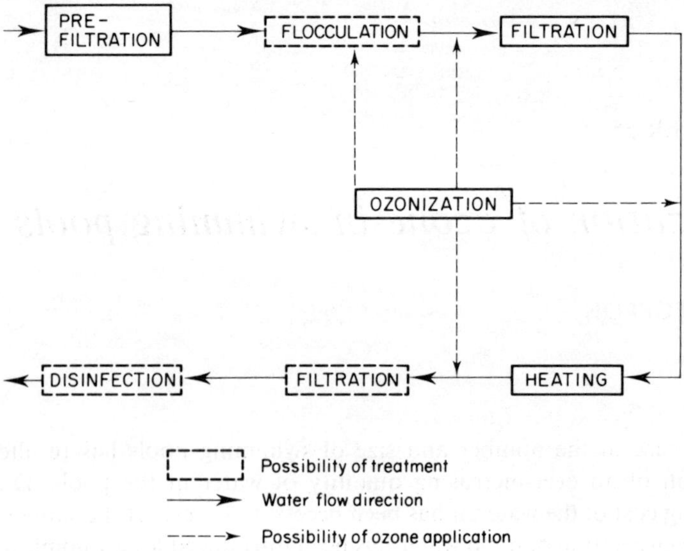

Figure 1. Ozone applications in the treatment of recycled waters of swimming pools

compartment circulates counterflow with ozonized air; this is where the oxidization reactions take place in order to meet the short-term ozone demand so as to obtain a residual of 0.4 g O_3/m^3 after a contact time of generally 2 minutes. Basically, the second compartment is used to maintain the residual concentration during 4 minutes: this is the stage of thorough disinfection. Finally, it is in the third section of the ozonization chamber that the residual ozone in water is eliminated: this can be achieved by any mechanical or physical–chemical means. The construction materials used for the ozonization chambers are stainless steel when the chambers are pressurized (small volume), and concrete for contacting columns at ordinary pressure (see Figure 2).

The application of the above-mentioned bactericide and virulicide conditions makes it possible to meet the health standards and in addition it improves significantly the physicochemical qualities of water: absence of taste and odour, excellent transparency, brightness, slightly bluish colour, etc.

This high-quality water must be brought to the swimming pool, whereas contaminated water has to be discharged in order to be treated. This requires a very well-adapted hydraulic circuit. In practice, two systems seem to be convenient: mixed circuits and reverse circuits (Figure 3). In the first case water comes from the lower section of the side walls and is discharged from the surface (50–70 % of the overall flowrate) and from the bottom of the pool at the same time. In the case of reverse circuits, water comes from the bottom of the pool and is discharged solely by overflow. In both configurations, surface water is

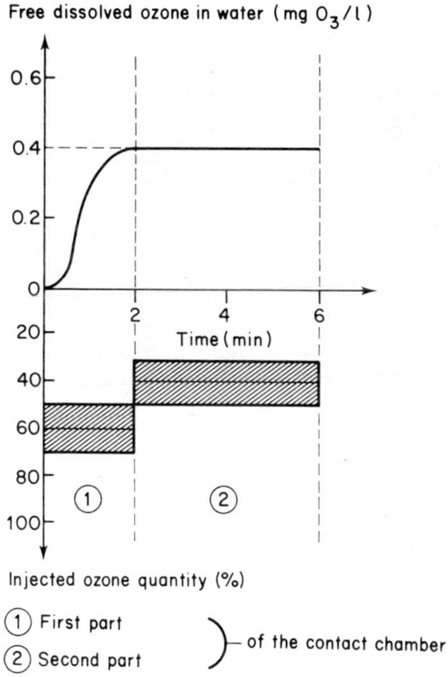

Figure 2. Ozonization conditions

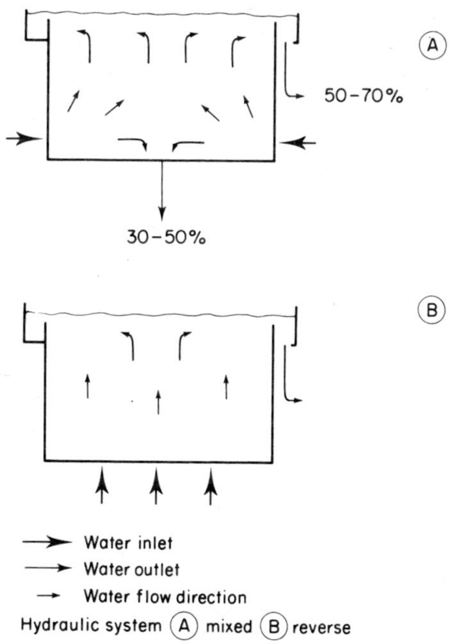

Figure 3. Recommended hydraulic systems

Figure 4. Part view of Olympic swimming-pool in Forbach (France)

continuously renewed, whereby the surface layer—often containing the higher germ concentration and floating material—is eliminated.

The residence delay of the water in the pool should be as short as possible, preferably not more than 4 hours.

The use of ozone in the treatment of recirculated water in swimming pools offers the advantages not only demonstrated by analysis but also claimed by the users themselves: the absence of inflammations or diseases, even after a long stay in the pool, is in favour of this treatment mode which, with or without addition of a longer remanence disinfecting agent, tends to be more and more in use.

FURTHER READING

G. Clerin, Réflexions sur six années d'exploitation d'une piscine traitée à l'ozone, International Ozone Institute Symposium, Los Angelès, 23–25 May (1978)

J. P. Legeron, Le traitement par l'ozone des eaux recyclées de piscines, 1er Congrès national du syndicat des directeurs de piscines centre sportifs et de loisirs, Pont à Mousson, 12 November (1977)

J. P. Legeron, P. Hartemann and F. Blech, Ozonation and swimming pool water quality, International Ozone Institute Symposium, Los Angelès, 23–25 May (1978)
J.-P. Legeron, Le traitement par l'ozone des eaux recyclées de piscines, Deuxième séminaire GRUTTEE sur l'ozone, Paris, 28–29 November (1978)
K. Orgler and J.-P. Legeron, Contribution de l'ozonation au traitement des eaux de piscines, 3rd Congress of the International Ozone Institute, Paris, 5 May (1977)

CHAPTER 56

The use of ozone in wastewater treatment

G. MARTIN and M. ELMGHARI-TABIB
CEEA–ENSEC Rennes

In the biological processes of wastewater purification, ozone can be applied at different stages. This oxidant will then come in contact with various types of materials: coloured soluble substances, suspension materials, colloidal substances, micro-organisms, molecular or polymolecular ions, for instance, humic acids in simple or complexed forms, colourless and non-ionized micropollutants, etc.

The polarized ozone can react with the substances by physical interactions. It can dimerize and chemically oxidize the reactives. This reaction may induce increased solubilization of the materials, biodegradability and so on; oxydation is usually fast and the oxidation by-products are not evidenced to be toxic for the environment.

These observations lead us to think that ozone may have different types of applications in secondary and tertiary wastewater treatment schemes.

SECONDARY TREATMENT

The injection of ozonized air into wastewater only slightly improves its BOD following a one-order reaction regarding BOD and ozone concentration. Therefore direct action does not seem valuable; on the other hand, ozonized air can be injected before the biological unit or in the aerating contactor. In fact, a synthetic effluent composed of ethanol, NH_4Cl, Na_2HPO_4 is purified by activated sludge with a load of 0.5 kg BOD day^{-1} kg^{-1} TVSM (total volatile suspended matter) with a purification yield of about 70 % for BOD (the oxygen dissolved is about 2 or 3 g/m^3. Ozone injection at more than 3.5 g of O_3/m^3 air induces biomass death. On the contrary, an amount of ozone around 1 mg/l increases the purification rate by 93–100 %. The measurements of the activity of the biomass through the ATP level (adenosine triphosphoric acid) indicate a slight increase. It is interesting to observe that the sludge settling is markedly improved by ozone injection, the Mohlman index is reduced: for instance, for different mass loads:

$$\frac{I_M \text{ without } O_3}{I_M \text{ with } O_3} = 1.3 \text{ to } 2.5$$

After ozonization of certain nitrogenated organic compounds, tests have indicated a better assimilation of organic nitrogen. In the case of urban wastewater, a fraction of COD is not easily eliminated and the increase in biodegradability of this fraction can be estimated at 20–30%. For the urban wastewater treated with oxygen, a slight increase in yield for an ozone amount up to 1 g/m^3 has been noticed.

In general, ozone concentration in the feed air of the activated sludge should be limited to 1 g/m^3 so as to avoid foam formation.

TERTIARY TREATMENT

World-wide a relatively large number of wastewater plants use ozone after secondary treatment. The main aim of the treatment is disinfection, i.e. a high performance in micro-organism removal. The disinfecting action of ozone has been well established and consequently use of the sterilization of wastewater is out of the question, a disinfection level has to be defined as sanitary objectives.

Actually, the bacterial contamination risk is generally considered as eliminated (with Salmonella as indicator), when the total coliform concentration is lowered by a factor of 10^4.

With the objective defined as such, one must seek to destroy micro-organisms bearing in mind that ozone may react with other reactive substances. As a consequence, the amount of ozone to be supplied varies considerably with the physicochemical characteristics of the water.

Ozonization of a secondary effluent lowers the content of suspended matter and flotation is often noticed. The origin of ozone efficiency on suspended matter may be related to the contact particles to ozone bubbles. The particle phase of a secondary effluent is more rapidly affected by ozonization than the dissolved phase. This observation does not exclude the possibility of ozone reactions with dissolved compounds through dissolved ozone molecules or through radicals resulting from decomposition. In fact, ozonization removes suspended matter by flotation and also by oxidation and solubilization. Flotation interpretation may be approached by observing a reduction in potential zeta achieved directly or through the flocculation, for instance, of decomplexed iron in the ferric state. The turbidity related to suspended matter is also reduced by ozonization.

Water discoloration by ozone is fast and often spectacular. Ozone reacts on dissolved or suspended colouring matters. The kinetics of ozone transfer in

Table 1

Type of pollution	Treatment rate g of O_3/m^3	Contact time (s)	Percentage removal	Authors
Suspended matter	20	—	80	Nebel (1972, 1976)
	20	—	9.1 to 35	Gardiner-Montgomery (1968)
	40	—	35	Gardiner-Montgomery (1968)
	11.67 to 17.62	800–1600	36.54 to 64.6	Croda Report (1976)
Turbidity	15	—	70 (J.U.)	Nebel (1972, 1976)
Colour	40	—	90	Nebel-Boucher (1967) Weingarden (1971)
	20	—	30	Nebel (1972, 1976)
COD	11.8 to 18	800–1600	22.1 to 32.65	Croda Report (1976)
	9.1	540	20	Gomella (1979)
BOD	80	3600	50	Carthy (1975)
	9.1	540	35	Gomella (1979)
TOC	15	—	4.3 to 18	Nebel (1972, 1976)
Cyanides	6	—	80	Nebel (1972, 1976)
Nitrites	60	—	80	Nebel (1972, 1976)
Phénols	10.8	—	93	Nebel (1972, 1976)
	25.6 to 28.8	780	77.45 to 89.51	Gardiner-Montgomery (1968)
	25	—	80	Gardiner-Montgomery (1968)
Surfactants (ABS)	30	—	65	Gardiner-Montgomery (1968)
	50	—	90 to 100	Gardiner-Montgomery (1968)
Organo-chlorinated biocides	20	300 to 1200	16.1 to 93	Gardiner-Montgomery (1968)

solution seems to be the limiting factor for the oxidation of molecules with conjugated multiple bonds.

The elimination of the dissolved ozone depends on the presence of soluble organic matter with molecular weights between 150 and 18,000. The ozone demand of dissolved organic matter is reduced if the mass load of activated sludge is lowered. The BOD and COD of the effluent are reduced just as the optical density at 254 nm, but the TOC shows only little variation. So dissolved oxidated products are created, without CO_2 production; pH does vary but remains close to neutral or unchanged value.

Some specific organic compounds have been followed during the ozonization of a secondary water. For phenols, 1 to 5 mg of O_3 are required per milligram of phenol from low concentrated solutions. This ratio is related to the initial concentration of phenol. Other data indicate 5 moles O_3 should be required per molecule of phenol for complete destruction. For surfactants, the ratio might change from 5 to 8 mg of O_3/mg ABS (index). Inorganic reductors in solution also consume ozone, for instance, cyanides, nitrites, and ammonia in some instances. Two observations have to be made:

(a) It is impossible to fix a treatment ratio and ozone residual for an inactivation of 4 decades in total coliforms if the water type is unknown. For instance, the treatment of a water containing 30 mg/l in BOD and SM, is achieved by a dosage of 6–11 mg/l, usually 8–10 mg/l with a contact time of about 20 min. The treatment rate also depends on the residual COD.
(b) A preliminary physicochemical treatment such as flocculation, filtration, and adsorption on active carbon prior to ozonization is valuable before completing sanitization.

For a secondary effluent it seems that an inactivation of 4 decades can be carried out with the following conditions:

Contact time	1560 s to 780 s
Ozone rate	0.5 g/m^3 to 11 g/m^3
Residual	0.1–0.3 g/m^3 to 0.5–0.6 g/m^3

For a tertiary effluent, for instance after fast filtration, the following conditions might be applied:

Contact time	780 s to 540 s
Ozone rate	7.5–8.5 g/m^3 to 6.5–10.8 g/m^3
Residual	0.6 g/m^3 to 0.5–0.6 g/m^3

An equal sanitation of the effluent is obtained after biological filtration ($\sim$10 g/m^3 BOD and $\sim$5 g/m^3 suspended matter) with an ozone consumption reduced by about 30% as compared to that of the secondary effluent.

In addition ozone forms the vent of tertiary treatment and can be introduced into the biological secondary epurator with a positive effect. However its concentration is best when limited to 1 g/m^3.

FURTHER READING

G. Martin, Séminaire du Gruttee, Paris, France (1979)
J. McCarthy, *Proceedings of 2nd International Symposium on Ozone Technology, Montréal*, p. 522 (1975)
C. G. Hewes and R. R. Davidson, *J. Am. Inst. Chem. Engrs.*, **71,** 141 (1971)
H. B. Ghan, C. L. Chen and R. P. Miele, Congrès IOI, Los Angeles, May (1978)

P. L. Boucher, *Proc. 22nd Industrial Waste Conf., Purdue University*, no. 126, p. 771 (1967)
M. J. Weingarden, Conference at Institute on Ozonation in Sewage Treatment, Milwaukee, Wisconsin (1971)
C. S. Wynn and B. S. Birk, EPA Technology, Washington, DC, January (1973)
C. Nebel, P. C. Unangst and R. D. Gottschling, *Revue*, Purdue University, Part I, no. 141, p. 1039; Part II, no. 141, p. 1056 (1972)
D. K. Gardiner and H. A. C. Montgomery, *Water and Waste Treatment*, **12,** 92 (1968)
Rapport CRODA, 25 October: Le traitement des eaux résiduaires par l'ozone (1976)
D. M. Drago, P. Cescon, R. Franche, *I.N.Q.*, no. 237, p. 19 (1974)
C. A. Beucher and D. W. Ryckman, *Proceedings of the 16th Industrial Conference*, Purdue University, p. 251 (1961)
C. Nebel, R. D. Gottschling, P. C. Unangst, H. J. O'Neill and G. V. Zintel, *Water and Sewage Works*: Part I, **123** (4), 76 (1976); Part II, **123,** (5), 82 (1976); Part III, **123** (6), 81 (1976)
J. P. Legeron, Séminaire Gruttee, Paris, France—Rapport sur la station Chino Basin USA (1979)
S. Gomella, Séminaire Gruttee, Paris, France (1979)
B. Hugues, M. Plissier and J. P. Torres, *L'eau et l'Industrie*, November, pp. 67–72 (1977)

CHAPTER 57

Some examples of ozone use in industrial wastewater treatment

P. LE TAREAU and J. M. ROVEL

Degrémont

GENERAL

Thanks to its exceptional oxidizing power, ozone may be used to destroy a large number of organic molecules dissolved in industrial effluents.

Indeed, it has been noted that in most of the cases the kinetics of oxidation by ozone remains sufficient, even when conditions are those normally prevailing in wastewaters, i.e. temperature close to ambient, very diluted medium. However, rarely is the oxidation process achieved to its limit and, most of the time, only a certain linkage or fraction of the molecule is affected. This does not eliminate completely the biological or chemical oxygen demands, but the molecule is sufficiently modified to have its impact in the natural medium changed. It was thus observed, for example, that some complex organic molecules, little biodegradable and even toxic, are broken down into much more easily biodegradable elements.

Since such actions are relatively specific, each case should be considered as a particular one; however, it must be kept in mind that economic considerations are the basic factor in selecting any treatment by ozonization. This is demonstrated by the examples given hereafter.

EXAMPLES OF APPLICATION

Oxidation of simple or complex cyanides

The free cyanides, which may or not be in equilibrium with metallic complexes, are among the substances having the highest immediate toxicity. For this reason, they must be oxidized into cyanates CNO^-; the oxidation reaction can even be completed until $CO_2 + N_2$ is obtained:

Provided pH value exceeds 11.5, the oxidizing agent traditionally used is either sodium hypochlorite NaClO or dissolved chlorine.

A substitution of chlorine derivatives by ozone is favoured by several authors. It appears that oxidation into cyanates is easily obtained when pH exceeds 9–9.5, the alkaline reagent cost being thus lower; the automatic control of the reaction is as simple as when chlorine is used.

For example in a typical effluent (such as rinsing from an electrodepositing production line); it is observed that to oxidize 1 kg of CN^- into CNO^-, reagents are required as:

(a) either 21 l of hypochlorite (150 g/l of active chlorine): approx. 2 US$/kg of cyanide; or
(b) 1.85 kg of ozone, i.e. approx. 1 US $/kg of cyanide.

The saving in alkaline product and in acid to neutralize the resulting excessive alkalinity may be figured as approx. 0.2 US$/kg of cyanide.

Despite a significant saving in the running costs, the use of ozone may be interesting only for very large detoxication plants. In effect, the depreciation time for an ozonization unit lasts much longer than for a sodium hypochlorite unit (including only one storage tank and one metering pump).

Effluents from the dye industries

Ozone was tested by many authors to remove the colour from the effluents from textile finishing works. Typical results of ozone efficiency when applied to effluents previously submitted to a biological treatment combined to a chemical flocculation are given by the curve in Figure 1.

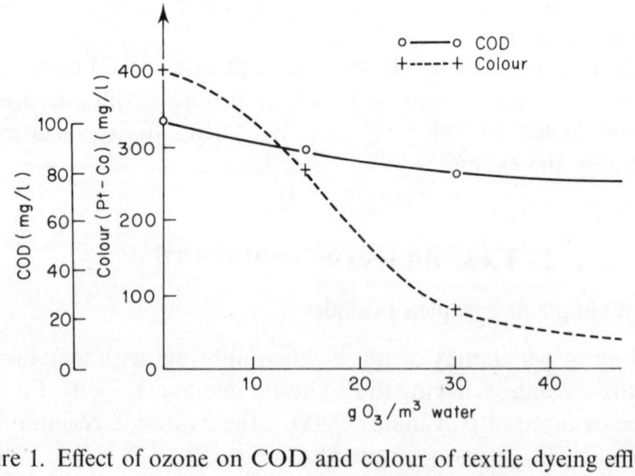

Figure 1. Effect of ozone on COD and colour of textile dyeing effluents

Provided rather extended contact times (30 min approx.) and rather low ozone concentrations are used, the colour removal effect is obtained preferentially to the oxidation of the other dissolved organic molecules remaining in the effluent.

In the described example, 25 g consumed ozone/m^3 correspond to 80% removed colour and only 20% eliminated COD.

In this particular case, the cost is similar to that of the other colour-removal treatments like activated carbon.

Other tests conducted on used dyebaths (metal-containing dyes) showed that colour removal was also feasible at this level. There again, the colour removal is 80–90% approx. for a COD removal limited to 10% approx.; it appears that the chromophoric groups of molecules are affected first (this principle does not necessarily apply to any type of dye).

In such cases ozone has a specific effect, e.g. colour removal on a previously treated effluent; the action of ozone is thus concentrated on the products which cannot be removed otherwise, and its high cost may thus be justified.

Effluents from oil-refineries

After a conventional physical–chemical treatment (primary oil removal followed by a coagulation–flotation), the effluents from a refinery still contain an essentially soluble pollution having typical characteristics as follows:

(a) COD	200–250 mg/l
(b) BOD	80–120 mg/l
(c) Suspended solids	15– 30 mg/l
(d) Hydrocarbons (IR)	5– 15 mg/l
(e) Phenols	1– 1.5 mg/l

Depending on discharge points, and on the various regulations applicable, such effluents may be discharged provided residual content either in phenol or in hydrocarbon removal is improved. Several studies were conducted in this direction:

(1) *Removal of phenols*

Phenols may be oxidized by ozone, down to contents not exceeding 0.2 mg/l by using about 7.5 kg ozone/kg removed phenol; this corresponds to running costs of about 50 US$/kg phenol.

Such costs are prohibitive. Indeed, for similar costs, a biological finishing process would remove not only phenols but also 80% BOD and 70% COD.

(2) *Hydrocarbons*

The action of ozone is different when essential pollutants are either, aliphatic hydrocarbons, whether or not saturated, or polycyclic or simple aromatic hydrocarbons. In the first case, ozone is practically inefficient. On the contrary, when aromatic hydrocarbons are concerned, ozone opens the aromatic nuclei and removes the factor dangerous to human health from such elements. For example, 3,4-benzopyrene (known as very carcinogenic), at 0.1 ppm concentration, is totally destroyed after an application of 1.5 g ozone/m^3 in a laboratory test where this component is present alone. On the other hand, the amount of ozone required is much higher to destroy it when numerous other competitively oxidizable substances are present.

Generally, the use of ozone in such cases is very difficult to forecast unless a previous pilot test has been conducted.

CONCLUSION

A large number of synthetic organic molecules may be destroyed through ozonization; oxidation may be completed more or less. However, owing to the complexity of industrial effluents, it is recommended to conduct previous tests in order to use ozone in the best possible conditions. Considering the high cost of an ozone production unit, it cannot yet economically compete with simpler treatments such as flocculation and phase separation or biological treatment, when concentrations exceed a few milligrams/litre and when applied alone instead of a treatment scheme involving a sequence of several phases.

Therefore at present, ozone may be mainly used within the following scope:

(a) specific treatments, such as colour removal and detoxication, when conventional treatments are inoperative;
(b) removal of the last traces before discharge to the natural medium or even re-use of the water.

It should be noted that in the latter case, the bactericidal and viricidal properties of ozone are also of great interest.

FURTHER READING

Anonymous, Verband der Chemischen Industrie, *Bericht 8*, p. 14 (1978)
Anonymous, *Galvano*, No. 363, p. 311 (1967)
Anonymous, Gruttee's 2nd conference, Exposé by Degrémont (1978)
Anonymous, *Brown-Boveri review*, p. 171, March (1978)
Anonymous, Gruttee's 2nd conference, Exposé by ELF (1978)
M. Mignot, Degrémont documents (1979)
Brooks and Shaw, *10A Congress, Paris (1977)*, IOA (1977)
M. Reichert-Mainz, *Water Sewage*, **110** (18), 969.
J. M. Rovel and P. Le Tareau, *Teintex*, no. 4 (1979)

CHAPTER 58

Possible use of ozone in papermills

F. A. ABADIE-MAUMERT

Norwegian Pulp and Paper Research Institute

Owing to the very good results obtained by ozonization of domestic wastewaters or effluents from different industries, it is obvious to take the use of ozone into consideration for the treatment of effluents from the pulp and paper industry. Especially in North America and the USSR, research has shown that the use of ozone presents interesting perspectives. However it is necessary to consider the following comments:

(a) Results published in technical literature are partially contradictory and often incomplete. Each mill has specific problems and the experience of a given company cannot be directly transposed to another, particularly when the costs of the treatment are simultaneous to the evolution in the ozone technology concerned. The price of electric energy increases constantly and, today, it is not possible to foresee a stabilization. Also, the price of energy varies noticeably from one country to another.
(b) The amount of ozone necessary to obtain a quality-objective depends on the kind of wastewaters, their dilution etc.

Most of the published work indicates that by ozonization of the wastewater from the pulp and paper industry, it is possible to considerably reduce the odour and the colour of these effluents, their TOC and COD. On the contrary the BOD often increases. However, it is possible to lower the BOD to an acceptable level with subsequent biological treatment, later on in the process.

The results concerning the sterilization of these wastewaters are more versatile. It appears that there are different threshold values for the quantities of ozone to be used, and that these quantities depend on the goal-values or quality-objectives.

It is obvious that by varying the quantity of ozone, it is possible to obtain a residue which can yield marketable products after fermentation, or no harmful residue at all. At the same time, it is also obvious that there is a maximum effect-level of ozone and that it is useless to increase the quantity of ozone above a given level in order to try to improve the result of the treatment.

A treatment of the effluents from the pulp industry by ozone degrades the organic sulphur-containing compounds. However, little is known of the possible action of ozone on the residual chlorinated compounds in the wastewater from the bleaching. Present knowledge about the treatment by ozone of the effluents from the pulp and paper industry enables us to foresee that this treatment will be more and more a progressive option for this industry. However, it will also require further important work in research and development.

Up to 1980 research laboratories have tried to solve the problems of pollution in the pulp and paper industry not by treatment of the effluents, but by the development of new methods to limit the pollution. Norwegian and Canadian work has led to the development of methods which make improvement of the properties of mechanical pulps by ozonization possible. By these methods, it is also possible to reduce the quantity of chemical pulp, e.g. chemical sulphite pulp which is known as a great source of pollution. The new techniques have already reached the stage of industrial pilot plant, in which newsprint has been produced on modern paper machines and newspapers printed on fast printing presses.

The economic and technical advantages are obvious. However, with treatment by ozone the pollution as measured possibly appears greater than when mechanical pulp is not treated. Therefore the total weight of pollution from the effluents remains questionable.

Other works, especially Norwegian and Japanese, point out that it is possible to partially, or even completely, substitute oxygen and ozone for the chlorinated compounds, in the bleaching of chemical pulps. The improvement where pollution is concerned is unquestionable. The chlorinated compounds in the effluents are eliminated or considerably reduced. The destruction of these compounds is very difficult and, as mentioned above, it does not seem possible to eliminate them by an ozonization of the wastewaters. Moreover, as the alkaline compounds used for the bleaching can be regenerated by combustion, there is an obvious reduction of the TOC, COD and often BOD in the wastewaters.

Thus the pulp and paper industry opens very interesting possibilities for ozone, needing considerable amounts of ozone. These needs can involve new developments of the ozone technology.

FURTHER READING

J. Dorica and A. Wong, *Pulp Paper Can.*, **80,** 65 (T80) (1979)
R. E. Reizin'sh and A. D. Tupireine, *Khim. Drev.* (*Riga*), **3,** 55 (1975)
N. Soteland, *Int. Mech. Pulping Conf.* (Helsinki), Proc., Vol. II, Paper 12 (1977)
N. Soteland, *Norsk Skogind*, **32,** 199 (1978)

CHAPTER 59

Ozonization of bottled water

W. SCHNEIDER

Fresnius Institut

Even the cleanest and most modern bottling techniques for drinking or mineral water (so-called aseptic procedures, membrane filtration, etc.) cannot prevent a few viable germs remaining in the water or being entrained by the air or the closure. Though the bottled water is completely unobjectionable in its bacteriological condition, secondary germ multiplication leads to high colony counts (10^5–10^6/ml), dependent on the duration and temperature of storage. A decrease will occur after a certain time, possibly caused by the inhibiting effect of the micro-organism's own products of metabolism; however it is known that objections have been raised by authorities because of too high colony counts.

This secondary multiplication of germs can only be prevented by filling sterilized water into sterilized vessels. Bottling under ozone protection has been successfully employed for that purpose by numerous bottling plants. In addition, ozone oxides possibly present organic trace substances which can impair the taste; the oxidation products are retained on activated carbon. Thus, the ozone treatment also results in a distinct improvement of the taste quality of the water.

Ozonization can be carried out in one or two stages, according to the manganese content. One-stage treatment is only possible with manganese contents below 0.02 mg Mn/l. In this case, the water is treated with 0.3–0.5 mg of ozone/l. Any iron present is thereby oxidized to the trivalent form and can be removed by filtration through a gravel filter. After passing this filter the water still contains sufficient residual ozone for bottling.

In the case of manganese contents above 0.02 mg/l a violet colour is imparted to the water by permanganate ion and a two-stage procedure must be employed. After a first ionization the water is filtered through a gravel filter for removing iron, and is subsequently passed through activated carbon, where heptavalent manganese is reduced to the tetravalent state and is retained. The activated carbon filter also removes not, or only partly, oxidized organic compounds. After this filtration the water is in unobjectionable bacteriological and chemical condition. However, a certain re-infection in the activated carbon filter and during the bottling process cannot be prevented as excess ozone is decomposed by the activated carbon. Therefore, a re-ozonization is required, for which 0.1–0.3 mg of ozone/l is sufficient. This ensures that any germs brought in secondarily are destroyed and that moreover the space between water surface and closure is sterilized by the ozone slightly outgassing from the water. According to all experience, water bottled in this way, not requiring any addition of carbon dioxide, remains free of germs as long as the closure remains tight.

The bottled water has a redox potential of about +700 to +900 mV, according to the ozone content. Ozone decomposes with a half-life of about 1 hour in the aqueous phase to molecular oxygen. Hence, this procedure does not bring any foreign matter into the water, as e.g. in the case of silver or chlorine treatment. No traces of dissolved ozone are detectable after about 8–10 hours. If however, somewhat higher redox potentials are measured and colour reagents seem to prove the presence of residual ozone, then this is caused by oxidation products (e.g. of bromide or iodide) and not really by ozone. However, this phenomenon is usually observed only in case of bottling without activated carbon filtration.

Ozone is more stable in the humid gaseous phase of the bottle, but also here it is completely decomposed after a few days and can then no longer be perceived by its odour. Therefore, taking into account this longer stability of ozone in the gaseous phase, a water bottled under ozone protection should not be delivered to the consumer earlier than 3 days after bottling. Earlier consumption is not harmful, but a slight unpleasant odour of ozone may be noticed when opening the bottle.

Glass bottles or ozone-resistant plastic bottles made of PVC or polyethylene may be used for bottling ozone-treated water. In case of plastic bottles special care has to be taken that no secondary impairment of the taste of the water (which is extremely sensitive in that regard) will be caused by traces of organic substances set free from the plastic material. Careful selection of a suitable plastic material is therefore of particular importance. This also applies to the plastic material used in the closures. Preliminary tests with the selected materials and with storage times up to 12 weeks under practice-like conditions are advisable. Bacteria-tight sealing is particularly important in case of filling ozonized water into plastic bags.

Table 1 shows a typical development of the colony counts of water with and without ozone treatment. The water was filled into glass bottles with screw caps and stored for the times stated in a standing position at 20°C in diffuse daylight. The initial ozone content was 0.2 mg/l.

Table 1.

Test made after a storage time of	Colony counts per ml after 48 hours of incubation on gelatin at 22°C: Blank (without ozone)	Test (with ozone)	Ozone content (mg/l)	Redox potential (+mV)
0.5 hours	50	1	0.2	900
3 hours	70	0	0.1	900
6 hours	100	0	0.05	850
1 day	150	0	n.d.	200
2 days	500	0	n.d.	200
7 days	1,500	0	—	—
14 days	10,000	0	—	—
28 days	40,000	0	—	—
100 days	3,000	0	—	—

n.d. = not detectable.

FURTHER READING

W. Schneider, Anforderungen an ein Wasser zur Abfüllung als Quelltrinkwasser und Gedanken zur Abfülltechnik. Das lebensmittel-Labor, Schwerpunkt, *Milchwirtschaft* (Beilage der Zeitschrift *Deutsche Milchwirtschaft*), **1**, 77 (1974)

W. Fresenius and W. Schneider, Bemerkungen zur keimfreien Abfüllung von Wasser. *Das Erfrischungsgetränk/MWZ*, **29**, 303 (1976)

8. OZONIZATION FLOWSHEETS

CHAPTER 60

Ozone at Nice

P. DECOURCELLE

Compagnie Générale des Eaux

The city of Nice is one of the first in the world to have used ozone for the disinfection of drinking water.

The first ozonization installation, constructed by the inventor Marius Paul Otto, was put in service in 1907 for a daily flow rate of 22,500 m^3 of surface water. The daily ozone production could reach 24 kg and the energy consumption was 24 Wh for a dosage of 0.9 g of ozone per cubic metre of water.

A second facility was put in service in 1909, for a daily flow rate of 13,000 m^3, and a third one in 1925 for another daily 13,000 m^3. These installations provided excellent water, without taste or colour, *never* bearing the slightest pathogenic germ.

In 1970 the Municipal Administration decided to replace the older works by a unique ozonization facility capable of treating 90,000 m^3 daily, which was put in service in 1972 at Super Rimiez (Figure 1).

The daily ozone production reaches 216 kg; energy consumption is 18 Wh for dosages that can attain 2.14 g of ozone per cubic metre of water, thus assuring perfect sterilization and complete inactivation of viruses.

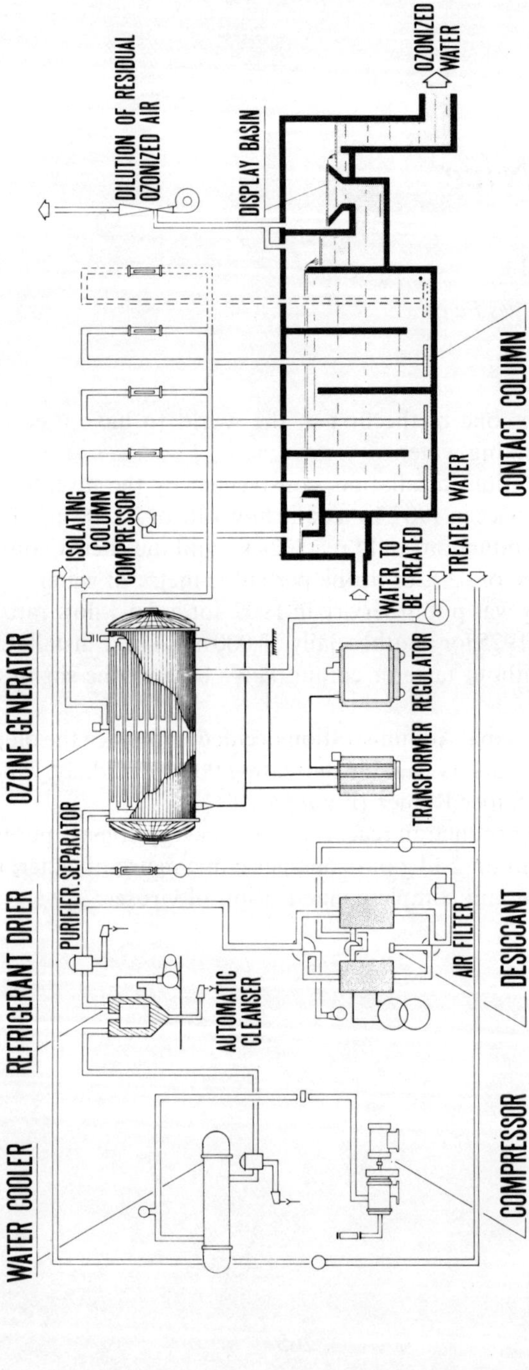

Figure 1. Diagram of the treatment plant at Super Rimiez

CHAPTER 61

The use of ozone in the treatment of water pumped from Lake Constance

D. MAIER

Betriebshabil- und Forschungslabor der Bodenseewasserrersorgung

The Zweckverbond Bodenseewasserversorgung has exploited since 1969 an ozonization phase in the treatment scheme of a plant using water from lake constance pumped at a depth of 60 m.

This plant can treat 27,000 m^3/h raw water for the obtention of drinking water. The global flowsheet is composed of five treatment sequences: microstraining, ozonization, reaction and detention, rapid sand filtration, and safety chlorination for water transportation.

The whole scheme is illustrated in Figure 1.

The raw water source of the plant is located at a depth of 60 m and 400 m from the lake bank. The ozone production capacity, as installed, enables the production of 32.4 kg O_3/h while the normal ozone dosing amounts to 0.9 g O_3/m^3. The ozone is put into contact with the water in packed towers and the action is completed in buffering or intermediate reservoirs of 70,000 m^3 each, thus providing for a contact time up to 2 to 5 hours depending on the filling height in the reservoir.

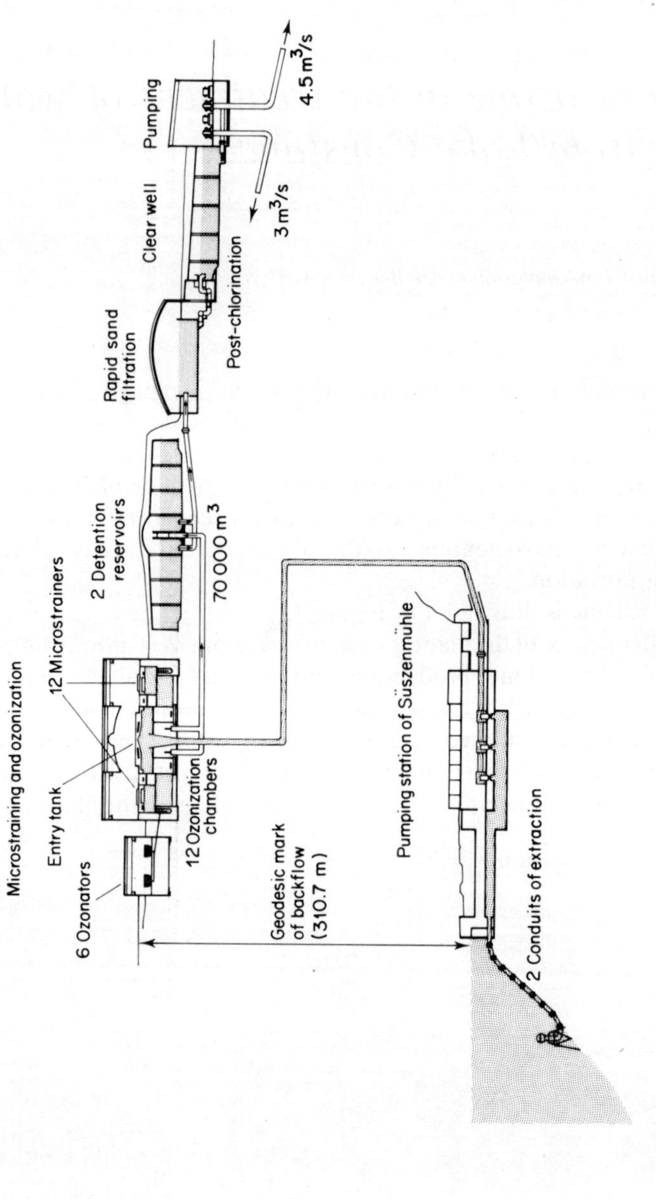

Figure 1. The flowsheet of the Sipplingen plant

CHAPTER 62

The ozonization scheme at the Tailfer plant

W. J. MASSCHELEIN

Compagnie Intercommunale Bruxelloise des Eaux

The Tailfer plant to treat water of the river Meuse by the CIBE applies ozone according to the concept of the post-ozonization as a finishing treatment. The unit processes involved in the flowsheet of the purification scheme as a whole are: pH adjustment; prechlorination; oxidation with chlorine dioxide; coagulation with an aluminium-containing reagent, this process being assisted through activated silica; introduction of powdered activation carbon followed by flocculation–settling in a sludge blanket clarifier, aggressivity correction with sodium hydroxide and intermediate chlorination followed by sedimentation, rapid sand filtration and ozonization.

At its final state before the transportation the water is still stabilized through a chlorination–post-ammoniation and a treatment with chlorine dioxide at the service reservoirs.

As a 1966–68 concept the flowsheet does not enable directly a preozonization step of the raw water but some of the favourable effects of this are obtained by the preoxidation with chlorine dioxide and the possible benefits of an eventual preozonization are at present under consideration and practical experience.

The ozonization as installed is performed by two consecutive injections during 5 min 22 s as a whole followed by a detention phase of 6 min 7 s. The functional criterion adopted is at all times to obtain a residual concentration between 0.3 and $0.5\,g\,O_3/m^3$ after the detention period.

The whole process is indicated by the joined cross-section in Figure 1.

An important particular aspect of this plant is the use of oxygen to enrich the process gas submitted to the electrical discharge. This aspect is discussed more in detail in Chapter 30.

FURTHER READING

A. F. Achten, *Technique de l'Eau*, **333**, 35 (1974)

W. J. Masschelein, *The Tailfer Plant of the Brussels' Intercommunale Waterboard* (ed. M. Slythoff) (1980)

W. J. Masschelein, *Congress IOA, Berlin, 1981*, Belgian experiences in water ozonization, O. H. Hess, Berlin, p. 434 (1981)

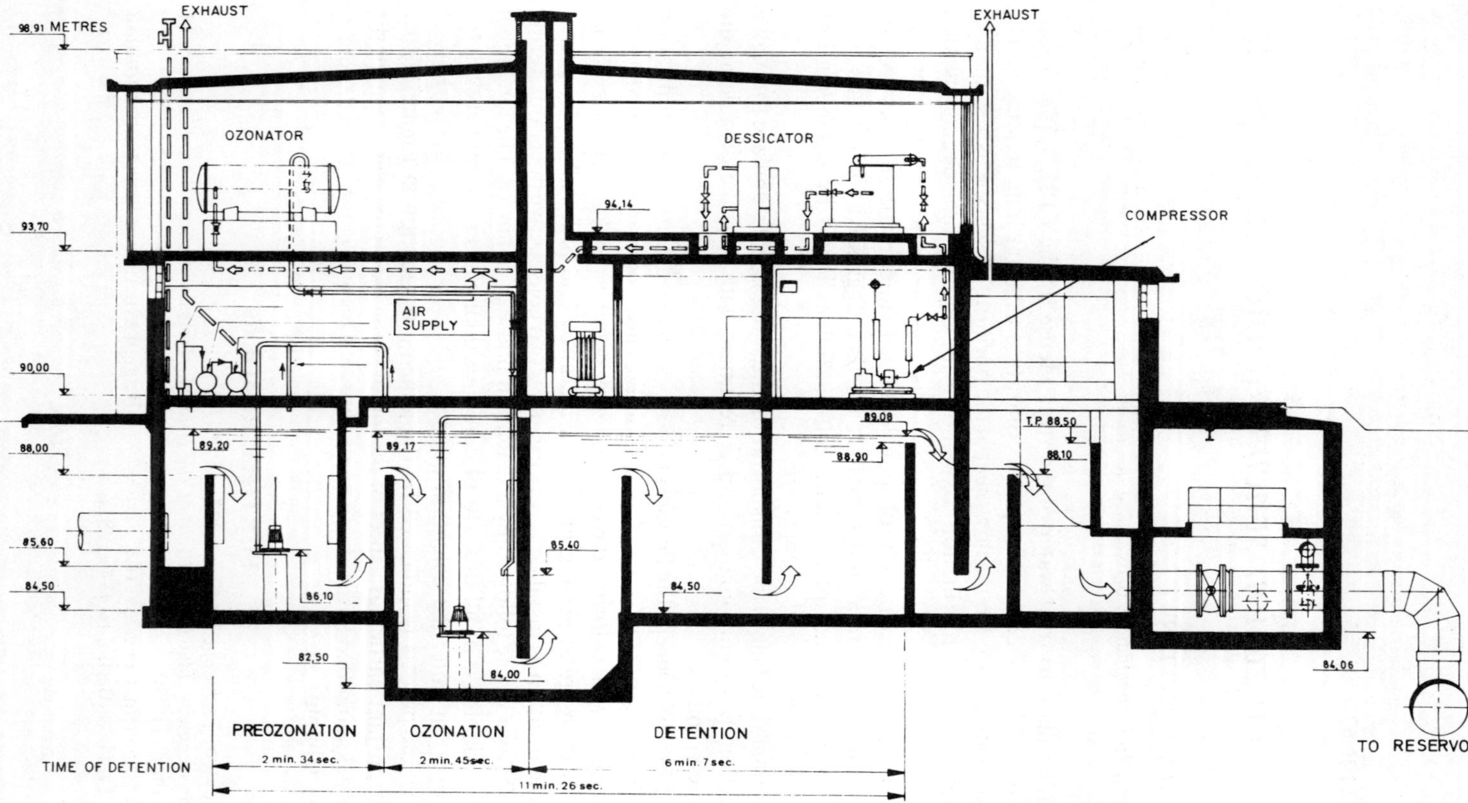

Figure 1. Flowsheet of the ozonization of the Tailfer plant

CHAPTER 63

Ozonization unit at the plant of Crissey (Chalon-sur-Saône)

J. MIGNOT

Degrémont

The plant of Crissey is basically designed to remove iron and manganese, and secondarily, to reduce the ammonia concentration. Indeed this 300 m^3/h capacity plant treats water which contains a high amount of iron (3 to 6 mg/l), manganese (approx. 0.15 mg/l), and ammonia (0.7 to 2 mg/l).

The treatment comprises cascade-type aeration in a closed building; the water falling over the steps is swept by residual ozonated air which comes from the ozonization chamber downstream from the cascade, and which represents a contact time of 7 minutes. The cohesion of the formed precipitates is increased by injection of alginate; these precipitates are next directly retained in open filters.

(a) The first two treatment phases, i.e., cascade aeration and ozonization, allow good water oxygenation, iron and manganese oxidation, and, which is exceptional, high ammonia removal. With pH 7, the ammonia content is reduced from 1.8 mg/l in the raw water to 0.3 mg/l on an average in the ozonated, aerated water: this reduction is probably due, for most part, to a stripping effect.

(b) Filtration takes place in open, dual-media filters, Mediazur T type, filled with anthracite (0.85 mm effective size) and sand (0.5 mm effective size). These units have a filtration rate of 6.80 m/h.

Excellent results are obtained with dosages ranging from 1.7 to 2.1 g/m^3 of ozone and 0.7 to 1 g/m^3 of alginate: manganese is totally removed from the filtered water; the ion content is reduced to less than 0.1 mg/l; the ammonia content remains the same as before filtration (approx. 0.3 mg/l). This is due to the fact that the presence of residual ozone in the feedwater to the filters inhibits the growth of nitrifying bacteria.

Final disinfection is done using a low amount of chlorine.

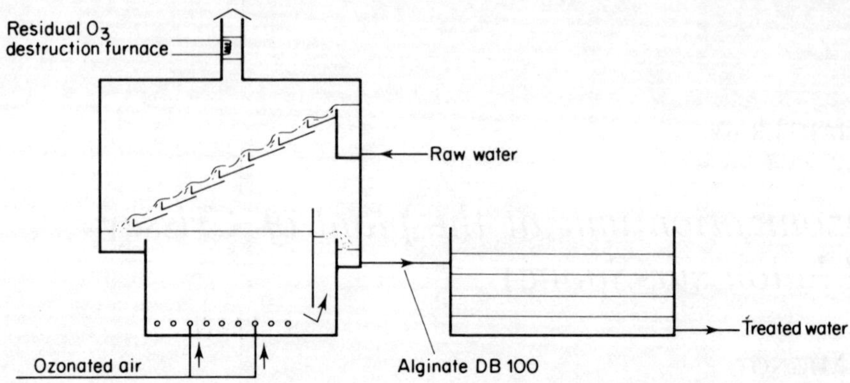

Figure 1. Flow diagram of Crissey plant (Chalon-sur-Saône)

CHAPTER 64

Plant of Anconella of the city of Florence

J. MIGNOT

Degrémont

The plant of Anconella handles a flowrate of 4 m^3/s of water diverted from the river Arno by a dam, then, after screening, pumped to the treatment structures. Ozone is applied for refining and disinfection (see Figure 1).

The treatment includes the following phases:

(a) Prechlorination using sodium hypochlorite and chlorine dioxide.
(b) The global water flow is divided into two schemes of different construction dates, clarification in scraper clariflocculators and open filters for 1.2 m^3/s, and in four Pulsator clarifiers and twelve Aquazur V filters for 2.5 m^3/s. The injection of powdered activated carbon during settling ensures the removal of compounds promoting bad taste.
(c) Ozonization for virus removal and general refining of the filtered water
(d) Final filtration through granular activated carbon, to be carried out later in case of pollution increase.

The ozonization unit produces 30 kg/h for the flowrate of 4 m^3/s: thus enabling a dosage of more than 2 g/m^3 which may be increased by adding another ozonator to the already installed three. Each ozonator has a capacity of 10 kg/h and is of the double-part type, and can thus operate with 275 tubes out of the 550.

They are fed with dry air with a dew point of -50°C; this air is produced by two drying lines comprising air compression, water cooling, passage through a refrigerating unit and polishing through an adsorbant. Ozonators are supplied with a.c. at variable voltage (8 to 20 kV), 50 Hz.

The ozonized air is injected into three concrete chambers ensuring a contact time of approx. 10 min at the maximum flowrate. Each chamber comprises two compartments passed through by a flow of water countercurrently to a flow of ozonized air injected by a system of porous tubes. The residual ozonized air at the outlet from the chambers is disposed of in a furnace equipped with thermostated electric resistors.

ANCONELLA PLANT FLOW DIAGRAM

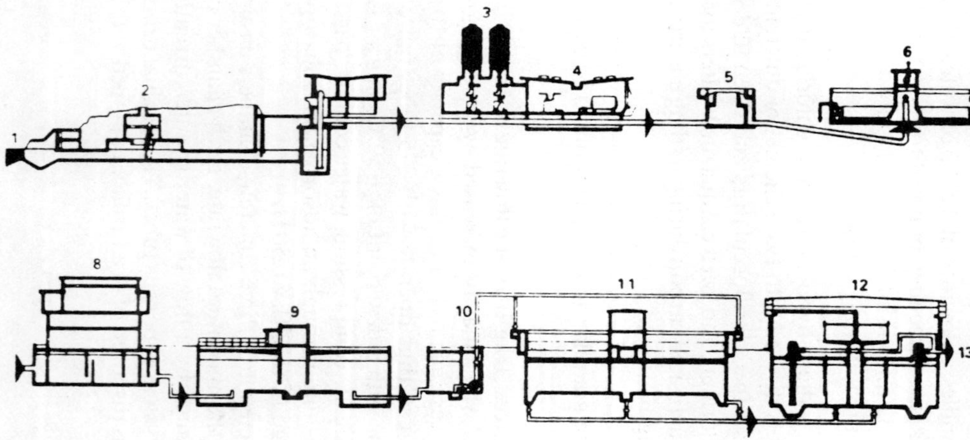

1. River Arno
2. Water intake & lifting
3. Lime & activated carbon silos
4. Chemical storage and metering
5. Distribution structure
6. Pulsator clarifiers
7. Aquazur filters
8. Ozone production unit
9. Make-up tank
10. Lifting
11. Activated carbon filters
12. Treated water pumping
13. Delivery to utilization points

Figure 1. Anconella flowsheet

The operation of the ozonization unit is entirely automatic; it is interlocked to the water flowrate and to the residual ozone content measured by an Amperazur detector at the outlet from the contact chambers.

The estimated dosage of ozone is displayed manually and automatically corrected by the electronic control system. This system also makes it possible to have a constant ozone concentration in air and to work under optimum conditions by automatically selecting the number and rotation speed of air compressors and the number of half-ozonators in service.

All control operations are displayed on an 'ozone' mimic diagram to allow both visual check of automatic operations and remote manual control of each element of the ozonization lines.

CHAPTER 65

Ozonization unit at the plant of Morsang-sur-Seine

J. MIGNOT

Degrémont

The Morsang plant was first designed to reinforce the facilities of the Société Lyonnaise des Eaux in the southern suburbs of Paris and to supply the new town of Evry with drinking water. At present, it is intended to remove micropollutants and to combine the use of ozone and activated carbon (see Figure 1).

This first stage, started up in 1970, is capable of handling 75,000 m^3 of raw water from the river Seine per day. The treatment includes prechlorination, flocculation–settling in a Pulsator-type sludge blanket clarifier with injection of

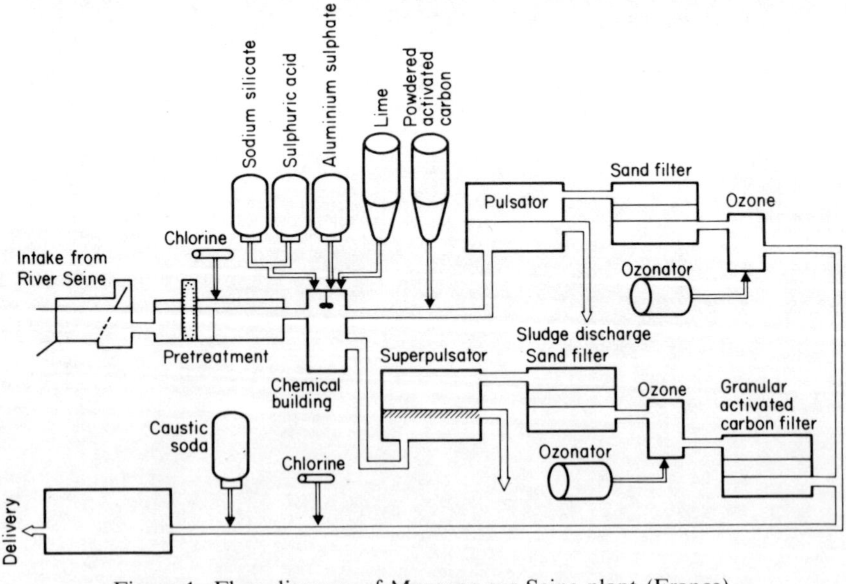

Figure 1. Flow diagram of Morsang-sur-Seine plant (France)

powdered activated carbon, filtration and ozonization. In this treatment scheme, ozone both improves filtered water regarding oxygen, detergent and organic matter content, colour and taste threshold, and ensures thorough disinfection by its viricidal action. The final injection of a low amount of chlorine ensures the presence of traces of residual disinfectant in the distribution network.

A second 75,000 m^3/day stage was commenced in 1975. The treatment was adapted to an increase in the pollution of the river Seine. After prechlorination, flocculation–settling in a Superpulsator-type sludge blanket clarifier (without injection of powdered activated carbon), and rapid filtration through sand, ozonization takes place, followed by filtration through granular activated carbon and final injection of a low amount of chlorine. Besides its above-mentioned effects, ozone also lengthens the service time of the granular activated carbon used in final filtration by favouring the bacteriological activity in activated carbon beds.

The dosage corresponding to the maximum flowrate is of 2.25 g ozone per cubic metre water. It is ensured by two ozonators, one per stage, each capable of supplying 7 kg/h from dry air (power supply: 50 Hz). The entire ozonization unit is fully automatic: ozone production is related to the treated flow and to the residual ozone content to be maintained at 0.4 g/m^3 in the second compartment of the contact chamber. The first compartment serves to meet ozone demand and reach the residual content of 0.4 ppm at the outlet.

This plant is scheduled to be extended until a maximum capacity of 1,000,000 m^3/day is reached in year 2000 to meet the water requirements of the new towns and urban centres of the southern and south-western suburbs of Paris.

CHAPTER 66

Ozonization unit at the plant of Roberval, Canada

J. MIGNOT
Degrémont

The Roberval plant, built in 1970, handles a flowrate of 9,000 m^3/day of water from Lake Saint-Jean; this water permanently features a high coloration (50 units) and an iron content of approx. 0.25 g/m^3. Turbidity occurs from time to time, thus requiring removal. Colour improvement and iron removal are the basic objectives.

The ozonization unit produces 41 kg/day, corresponding to a maximum available dosage of 4.5 g/m^3. Ozone–water contact takes place in two contact chambers, one upstream and the other downstream of the filtration unit.

(a) When the water has a low turbidity 66 % of the ozone amount is used for the pretreatment, i.e. a maximum dosage of 3 g/m^3, to reduce the colour and

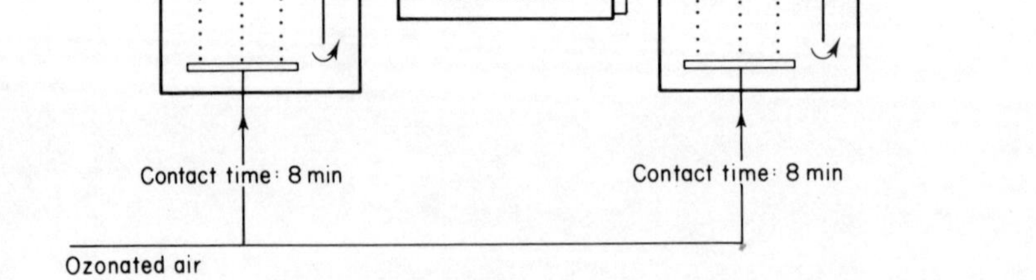

Figure 1. Flow diagram of the Roberval (Canada) plant

oxidize the iron before filtration; the remaining third is available for final disinfection. Thus, a colour of between 40 and 60 units in raw water is reduced to 10–15 units in pre-ozonized water and to 5 units after post-ozonization.

(b) During the snow thawing period, when the water of Lake Saint-Jean becomes turbid, a low amount of aluminium sulphate must be added after pre-ozonization so as to help flocculation of the colloids and their retention in filters.

CHAPTER 67

Ozonization at the treatment plants of the 'Syndicat des Communes de la Banlieue de Paris pour les Eaux'

M. RAPINAT

Compagnie Générale des Eaux

The Syndicat des Communes de la Banlieu de Paris disposes of three treatment plants supplied with river water and the design capacity as indicated hereafter:

(a) Choisy-le-Roi (900,000 m^3/day) on the Seine
(b) Neuilly-sur-Marne (600,000 m^3/day) on the Marne
(c) Méry-sur-Oise (270,000 m^3/day) on the Oise

The raw water is similar at all three plants in that it is very turbid, heavily loaded with ammonia (1 to 5 mg/l), and very rich in organic matter (TOC: 10 mg/l). The pollution levels of the rivers, already substantial on the average, can be considerably increased by accidental pollution.

Until now the three plants were set up with chemical treatment schemes as breakpoint chlorination, flocculation–decantation with the addition of powdered activated carbon, rapid sand filtration and viricidal ozonization at the final stage. The ozone production capabilities installed are:

(a) Choisy-le-Roi 2.880 kg 03/day
(b) Neuilly-sur-Marne 2.880 kg 03/day in four units
(c) Méry-sur-Oise 1.570 kg 03/day

In order to limit the formation of organic chlorine compounds and other by-products, it was decided to operate a biological nitrification treatment. The new treatment is installed at Méry-sur-Oise and comprises three successive ozonization stages:

(a) a pre-ozonization before raw water storage where the objective is to ameliorate the natural phenomenon of decantation and the elimination of

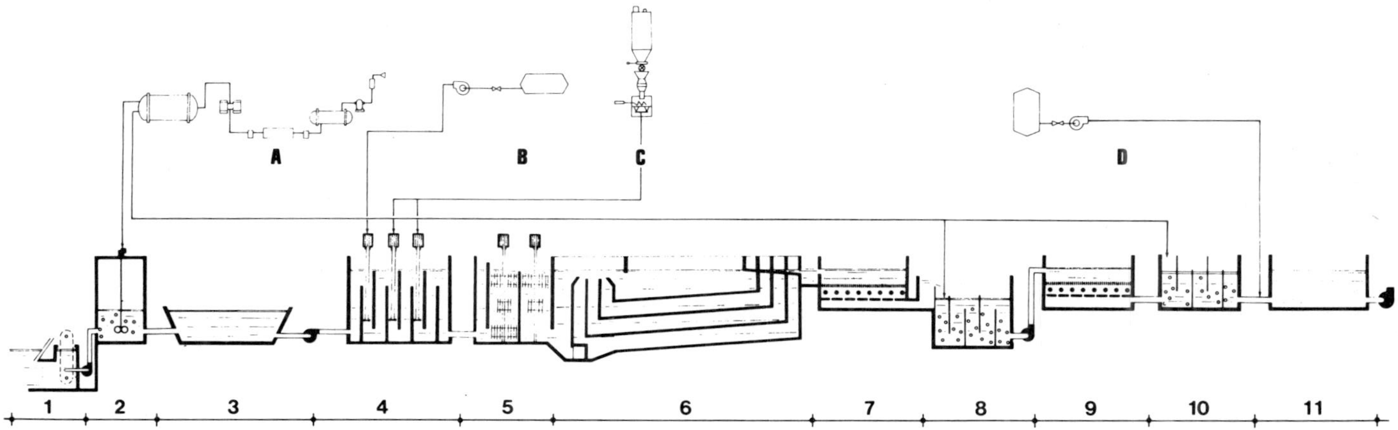

Figure 1. Diagram of the new treatment line of the Méry-sur-Oise plant

A. Ozone
B. Aluminium polychloride
C. Powdered activated carbon
D. Sodium hypochloride

1. Water from the Oise river
2. Ozonization
3. Raw water basin
4. Chemical treatment
5. Flocculation
6. Sedimentation
7. Sand filtration
8. Ozonization
9. Activated carbon filtration
10. Ozonization
11. Storage of treated water

organic matter that takes place during storage (capacity 170 kg 03/day);

(b) an ozonization before filtration on granular activated carbon—this second stage is operated to increase the elimination of some of the organic matter by biological processes in the carbon;

(c) a third ozonization after carbon filtration at viricidal rates to achieve purification and disinfection of the water.

CHAPTER 68

Ozone for biological nitrification and finishing treatment of groundwater at Rouen-La-Chapelle

J. CAUJOLLE-GAZET

Compagnie Générale des Eaux

The 50,000 m^3/day La Chapelle treatment plant, near Rouen, is one of the earliest biological carbon water treatment facilities. It was specifically designed for removing ammonia and organic micropollutants (Figure 1).

After a pilot investigation during 3 years, a process was developed for treatment of well waters drawn near the Seine which contain up to 3 mg/l of ammonium ion, 0.2 mg/l of manganese, various micropollutants (detergents, phenols, substances extractable with chloroform, etc.) and are practically free of dissolved oxygen. This process includes:

(a) pre-ozonization (0.7 mg/l);
(b) filtration through 1 m of quartz sand;
(c) filtration through 0.75 m of granular activated carbon (both filtration stages operate at 5 m/h);
(d) ozonization (1.4 mg/l);
(e) post-chlorination (0.1 to 0.2 mg/l).

Each treatment phase performs a well-defined role among which:

Pre-ozonization guarantees:

(a) Oxidation of iron, manganese, organics.
(b) Reduction of the concentration of various micropollutants, by oxidation and flocculation.
(c) Increase of dissolved oxygen, essential for efficient biological nitrification.
(d) Complex, biorefractory molecules are broken down and become biodegradable.

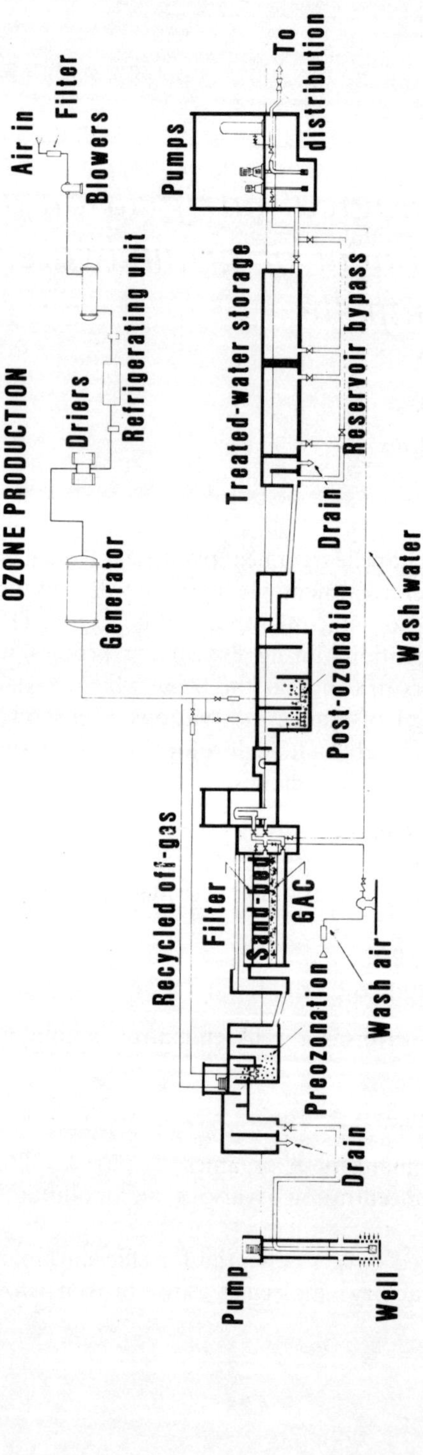

Figure 1. Treatment of the Rouen-la-Chapelle plant

(e) Conversion of soluble manganese and soluble iron to insoluble oxides, so they can be removed by the sand filter.
(f) Elimination of possible biological inhibitors which might interfere in later stages.

Post-ozonization, after filtration through sand and GAC, guarantees:

(a) Oxidation of micropollutants.
(b) Disinfection.
(c) Improvement of organoleptic quality.

Much of the ozone used in this first step derives from post-ozonization off-gases which have been recompressed and sent to a pre-ozonization turbine contactor.

Ozone is produced by three water-cooled generators, one of which is for backup and/or expansion (total production: 5.7 kg/h). Each ozone generator has excess generating capacity and is normally operated at 50 % of capacity, enabling lower operating voltages, maximizing power efficiency and minimizing stresses on the dielectric media.

FURTHER READING

R. G. Rice, C. Gomella and G. W. Miller, Rouen, France water treatment plant: good organics and ammonia removal with no need to regenerate carbon beds, *Civil Engineering—ASCE*, May, p. 76 (1978)

Environmental Protection Agency (EPA), *An assessment of ozone and chlorine dioxide technologies for treatment of municipal water supplies*, by G. W. Miller, R. G. Rice, C. M. Robson *et al.*, Cincinati, EPA, August, 600/2 (1978)

F. Damez and J. C. Dernaucourt, L'ozone avant flocculation, *La Technique de l'Eau et de l'Assainissement*, **388**, 17 (1979)

C. Gomella and D. Versanne, Biological nitrification and finishing treatment of a ground water: the performance data of the La Chappelle treatment plant after four years of operation, *Techniques et Sciences Municipales*, **75**, 211 (1980)

J. C. Dernaucourt, Developments in the techniques of biological nitrogen elimination: new constructions and prospects based on the experiments carried out at La Chapelle Works, *Techniques et Sciences Municipales*, **75**, 329 (1980)

CHAPTER 69

Ozone at the Annet-sur-Marne treatment plant

A. TIRET
Société Française de Distribution d'Eau

The Annet-sur-Marne treatment plant supplies drinking water to the Eastern suburbs of Paris. It was designed in such a way that it could be easily extended. The first stage, representing 25,000 m^3/day, was started in 1973. In 1977 a second stage enlarged the treatment capacity to 50,000 m^3/day. The treatment line in these first two stages included pre-oxidation with chlorine, clarification with 'Fluorapid' clarifiers acting both as flocculators and settling tanks, filtration, and final disinfection with ozone. In addition to its bactericidal and viricidal properties, ozone also helps to destroy some of the micropollutants in water, such as pesticides, phenols and detergents.

A third stage was put into operation in 1980, bringing the plant's total output to 70,000 m^3/day. The new phase comprises pre-oxidation using ozone, rather than chlorine or chlorine derivatives. This modification offers three advantages: less haloform formation, improved coagulation with saving in reagent, and increased bacterial activity for more efficient subsequent biological filtration, clarification; biological filtration with the 'Biocarbone' process; and final sterilization with ozone. The Biocarbone filters work with countercurrent insufflation of air into a granular activated carbon bed at a level about one-third from the filter bottom. This creates biological activity in the upper aerated area, with finishing filtration following in the bottom of the filter. During the major part of the year when the flow is not too high, the pre-ozonized raw water is admitted directly on to the Biocarbone filters without previous clarification, which allows us to decrease the need in chemical reagents considerably.

Ozone production at Annet-sur-Marne is assured by five ozonators with a total capacity of 12,150 kg/h.

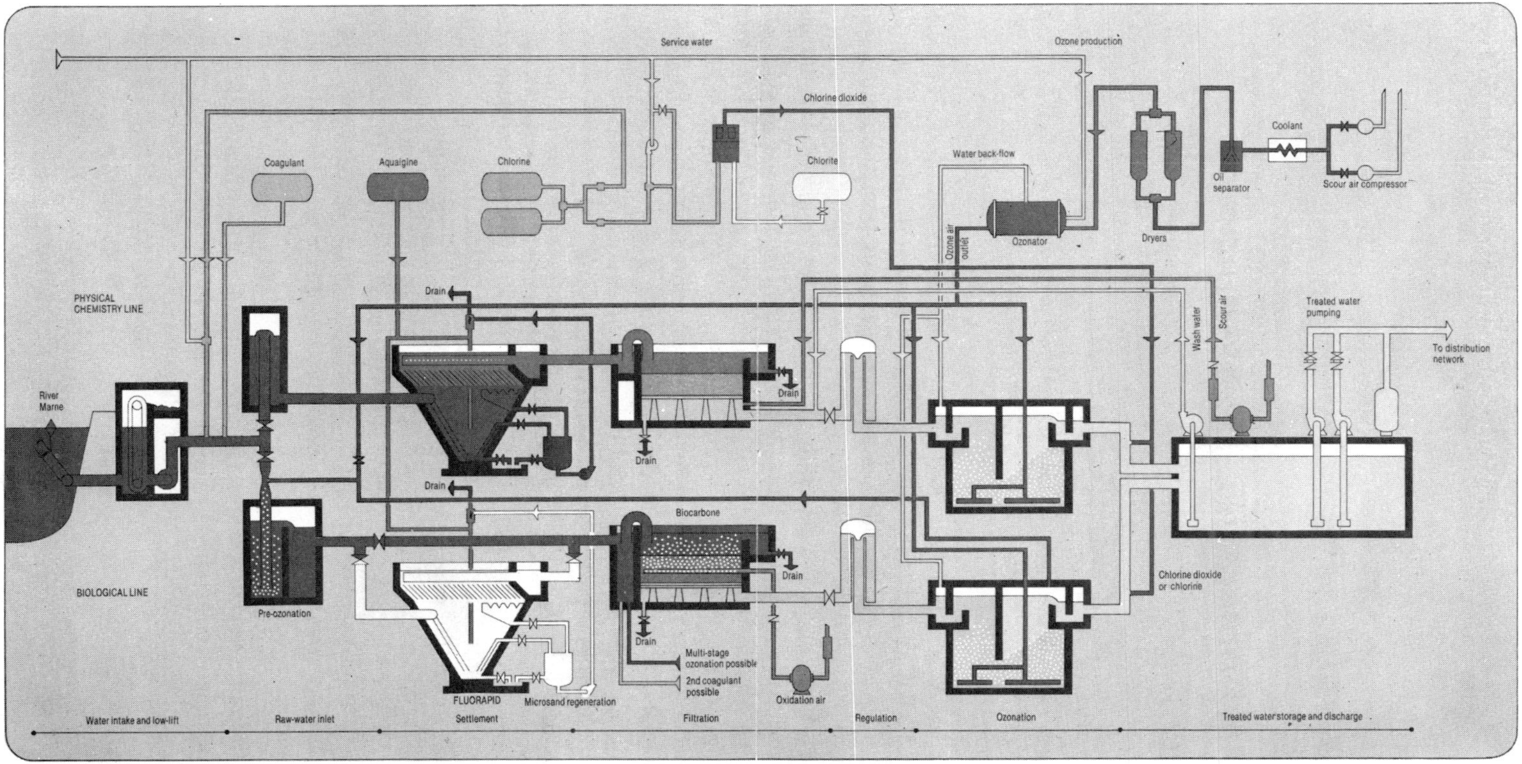

Figure 1. Hydraulic diagram

CHAPTER 70

Ozonization at the Ile-au-Bourg plant at Angers

J. DE FRESCHEVILLE

Compagnie Générale des Eaux

The city of Angers produces its drinking water partly with water of the River Loire, which is treated at the rate of 120,000 m^3/day at the plant of Ile-au-Bourg and with water from the alluvial stratum of the Loire pumped at rates varying seasonally from 40,000 to 80,000 m^3/day.

Water from the Loire generally contains 3 to 7 g/m^3 of organic matter oxidizable with permangate, 1 g/m^3 of iron, 150 to 700 million algal cells per litre and traces of ammonia. It currently undergoes conventional treatment: perchlorination, coagulation, clarification through Fluorapid filters, filtration, and final chlorination. Water from the stratum contains on the average 0.3 mg/l of manganese, 0.6 mg/l of iron and traces of ammonia. At present, it is distributed with no treatment other than final chlorination.

Ozone facilities are at present under construction.

Water from the River Loire will undergo two stages of ozonization:

(a) The first, pre-ozonization of raw water at 0.5 to 1 g per m^3 for destruction of algae and conditioning of the water for clarification. Tests have shown that pre-ozonization could result in a 17% saving in coagulants.

(b) The second ozonization between the Fluorapid clarifiers and the sand filters at an overall rate of 3 g/m^3 between the two stages. The aim is viral inactivation, preparation and the water for biological treatment in the sand filters and oxidation, which will reduce the chlorine demand at the end of the treatment process.

Water from the stratum will undergo an ozonization prior to sand filtration with the same objectives as for the second ozonization of Loire water, and also the oxidation of manganese contained in the alluvial water.

CHAPTER 71

Ozone at Rennes

J. DE FRESCHEVILLE

Compagnie Générale des Eaux

The city of Rennes first used ozone in 1963 in a plant treating water from the dam on the Rance. In 1975, ozone was incorporated in the city's new surface water treatment plant at Villejean, and in 1977, in a plant dating back to 1882 treating water from the local Minette and the Couesnon Rivers.

All three ozonization facilities, including the first, which was updated and extended in 1977, are designed for ozone production enabling viricidal conditions: to ensure a 0.4 g/m^3 residual action for ozone demands varying from 2.6 to 3.6 g/m^3.

All of the ozone generators operate at low frequency (50 Hz) and are fitted with tubes of 15 g/h unit production capacity.

Water from the very first supply, of which maximum flow is 1000 m^3/h, is treated in a facility of 3.6 g O_3/m^3 at nominal capacity. The ozone production is operated on a computer-controlled basis integrating the dominant parameters as water flow *vs.* residual ozone in the treated water. The ozone input is regulated as a function of the production controlled by the gas feed-rate of the compressors, the electrical power dissipated into the discharge of the ozonators and the residual concentration to be ensured in the water at its final treatment phase (Figure 1).

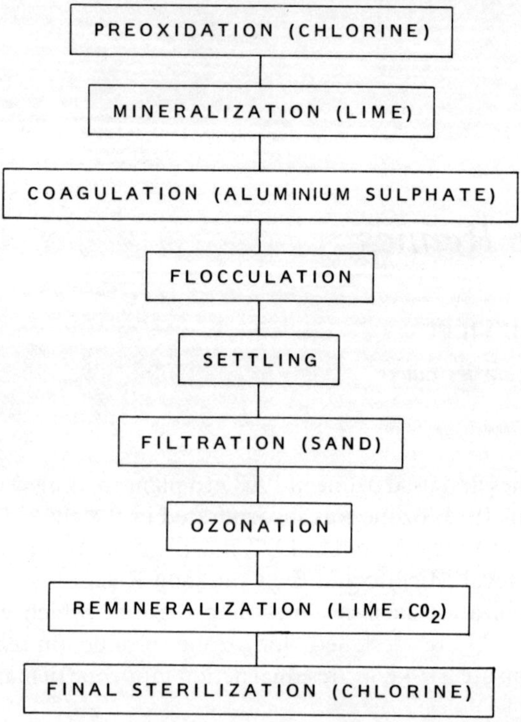

Figure 1. Diagram of the treatment line

CHAPTER 72

The Mauboule treatment plant at Valence

O. PASCAL

Compagnie Générale des Eaux

The city of Valence is supplied by water from various underground sources which are extracted at a mean rate of 26,700 m^3/day, either from galleries (7,400 m^3/day at Chabeuil) or from wells, as at Thabor (5,300 m^3/day), Couleurs (7,850 m^3/day) and Mauboule (6,150 m^3/day) (Figure 1).

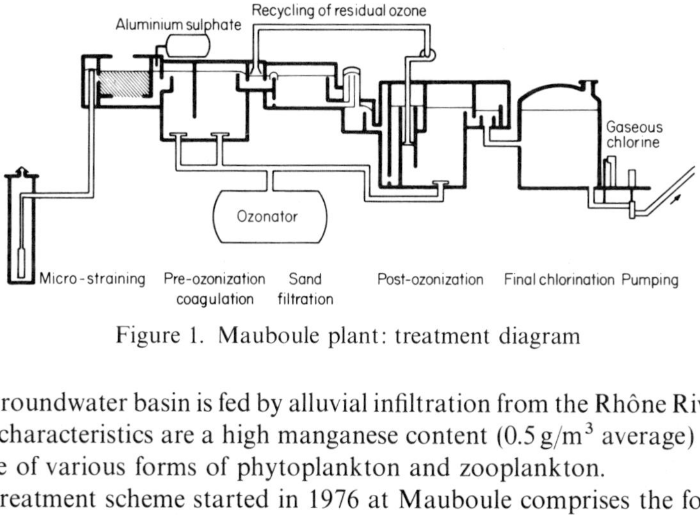

Figure 1. Mauboule plant: treatment diagram

The groundwater basin is fed by alluvial infiltration from the Rhône River. The water's characteristics are a high manganese content (0.5 g/m^3 average) and the presence of various forms of phytoplankton and zooplankton.

The treatment scheme started in 1976 at Mauboule comprises the following stages:

(a) *Water intake:* with eight groups of electric pumps; total capacity: 850 m^3/h.
(b) *Micro-straining:* with a 35 μm screen which eliminates most of the plankton and suspended matter, thus diminishing ozone demand as well as the frequency of back-washing of the filters.
(c) *Ozonization:* with a mean dosage of 0.5 g/m^3 in a 65 m^3 tank and a contacting time of 6 min. Ceramic porous tubes diffuse the ozonized air so as to ensure

oxidation of the manganese. Its precipitation is improved by coagulation with aluminium sulphate injected at the entrance of the contact columns. Also the elimination of non-filtrable planktonic cells is promoted.

(d) *Filtration:* through 0.90 m of sand; four filters with a surface area of 32 m^3 each operate at a maximum speed of 5 m/h. Filtration removes colloids the coagulation of which has been facilitated through ozonization.

(e) *Post-ozonization:* at dosages of 0.6 g/m^3 with a contacting time of 8 min. Porous tubes diffuse residual ozone from the first ozonization (recycling) together with fresh ozonized air. Final ozonization ensures complete disinfection of the water and viral inactivation.

Ozone production is provided by an 88-tube ozonator with a daily capacity of 32 kg of ozone, or a possible total dosage of 2 g/m^3 at nominal flow. The main feature of the Mauboule treatment plant is the two-stage ozonization, which ensures separate oxidation of manganese and viricidal sterilization. In a general way, a double-stage process improves, at a given overall dose rate, the efficiency of ozone's action.

CHAPTER 73

The new ozonization facility at Marseille

J.-P. LESBROS

Société des Eaux de Marseille

Since 1849, the city of Marseille has been supplied with water through the Canal of Marseille from the Durance River. Originally, the canal's source was in the river, but since 1963, it has been supplied with water deriving from the hydroelectric plant of Saint-Esteve (5.4 MW). The canal, about 100 km long and of a capacity of 17 m^3/s, is subject to a very severe control. Its management is assured by the Société des Eaux de Marseille in the context of a contract with the city of Marseille.

In 1973, a second complementary supply: the North-Marseille branch of the Canal of Provence, was put into service. With a capacity of 3.5 m^3/s, it transports water from the Verdon River to the reserve of Vallon Dol (3,000,000 m^3), situated above the city and acting as a security reserve.

The Canal of Marseille supplies the old drinking water treatment plants of Sainte Marthe and of Saint Barnabe by gravity flow. In 1974, the city decided to add a third plant, which was built near the Vallon Dol reserve. Ozonization is used in this plant, which is tele-controlled by permanent staff from the Sainte Marthe plant.

These three treatment plants supply, by gravity flow, most of the city of Marseille. The distribution is divided into three zones of different levels: (a) Sainte Marthe level 1.89 m (above sea level); (b) Saint Barnabe level 2.137 m; (c) Vallon Dol level 3.190 m.

The suburban distribution zones, at levels of 4 and 5 m, are supplied by pumping.

The Sainte Marthe plant (see Figures 1 and 2), which is the oldest, dates from 1934. Its current maximum production capacity—420,000 m^3/day—has been obtained by successive improvements. The newer plants were designed on the basis of the experience acquired in the Sainte Marthe plant.

The ozonization facility which has been installed at Vallon Dol was designed with future needs in mind, with an intermediate phase for 300,000 m^3/day,

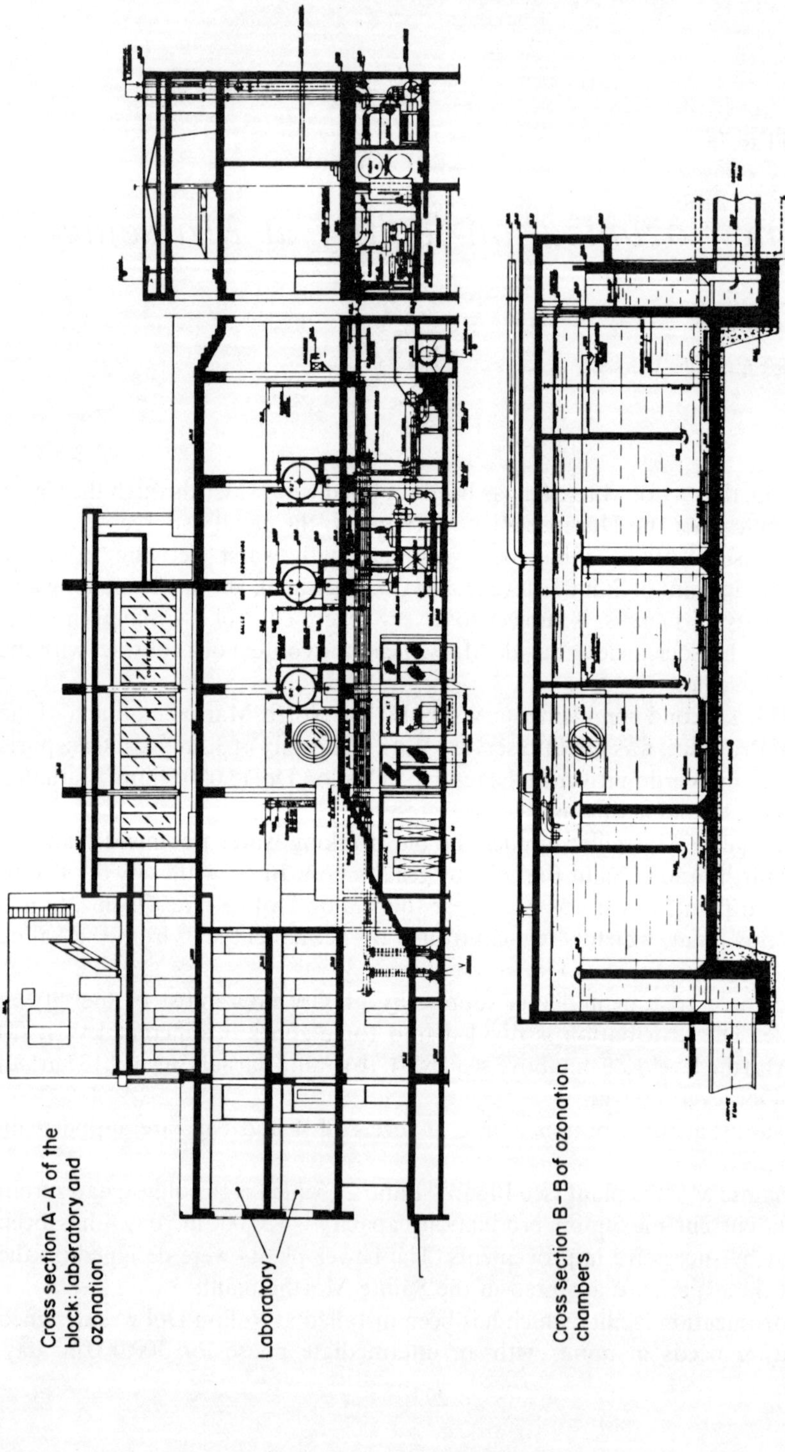

Figure 1. Treatment process at Ste Marthe: ozonization complex and laboratory (final phase)

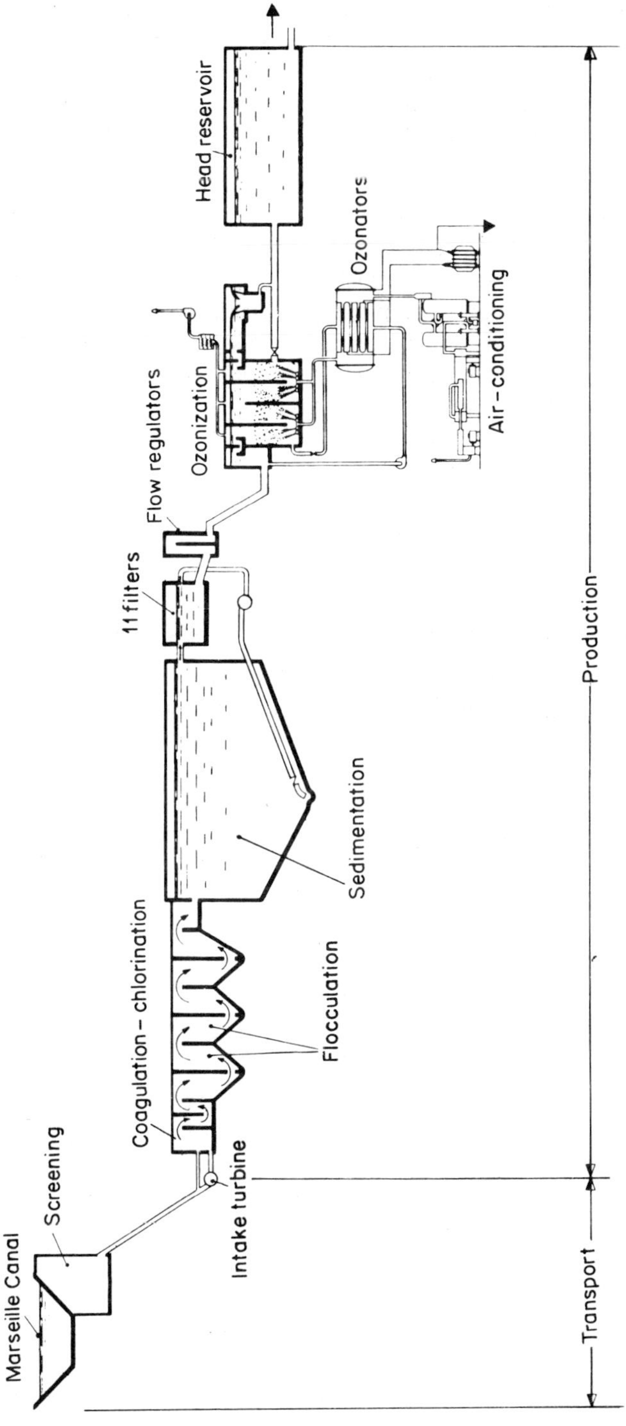

Figure 2. Facility at Ste Marthe: flow diagrams

corresponding to the flow currently treated. (Long-term needs for the Marseille area are estimated to be 520,000 m^3/day.) This ozonization facility was constructed and integrated in the new laboratories of the Société des Eaux de Marseille. The supply equipment and the auxiliaries are located in the basement and the ozonators on the first floor.

(a) The ozone production is designed for a regular treatment dosage of 0.7 to 1.5 g/m^3 and for 0.9 to 1.8 g/m^3 in the final phase, considering the changes of the ozone demand which may arise. The range of ozone production is thus 8 to 20 kg/h in the first phase and 15 to 39 kg/h in the final phase. To ensure both security and a good output, it was decided to install ozonizers capable of producing 24 kg/h: three ozonizers of 8 kg/h in the first phase, and to be multiplied later on to reach a 40 kg/h production capacity.
(b) In the same scheme, the contacting tanks were foreseen for an eventual flow rate of 520,000 m^3/day with a contacting time of about 8 min. In the first stage, a single line can treat the current flow with a contact time of about 7 min, and residual ozone in the off-gas is destroyed with a thermal destructor located at the end of the building. A circular reservoir of 12,000 m^3, put into operation in 1980, receives the ozonized water prior to its delivery to the distribution networks.
(c) The whole of the ozonization facilities is tele-controlled.

Once the ozonization facilities foreseen at the Saint Barnabe plant for 1982 are put into operation, all of the water purveyed in Marseille will be treated by this process.

CHAPTER 74

Application of ozone in the treatment of the Zürich lake water

M. SCHALEKAMP

Wassernersorgung Zürich

Since 1871, the Zürich Waterworks, which supply the city of Zürich as well as 57 suburbs (that is 750,000 inhabitants in all) with water, treats water from lake Zürich to produce drinking water.

Zürich actually has two lake-waterworks, one on the north bank of the lake: the Lengg plant with a capacity of 250,000 m^3/day and the other on the south bank of the lake: the Moos plant with a capacity of 120,000 m^3/day. The water intakes of the two plants are situated at about 600 m from the lake bank at a depth of about 32 m below the average lake level and at 16 m from the bottom.

Since 1967, Zürich has installed an ozonization phase in the treatment of the lake water.

A water-processing plant must supply pure drinking water at all times even if a serious problem of pollution should appear. Therefore Zürich uses an eight-stage water treatment system comprising pre-oxidation, micro-flocculation, neutralization, double-layer rapid-filtration, ozonization, GAC filtration, slow sand filtration and a chlorine dioxide treatment (Figures 1 and 2).

The proportions of different raw water sources are the following: 70 % Zürich lake water, 10 % spring water, 20 % groundwater.

The *spring water* is mixed with the lake water and treated in Moos (Figure 2).

The *groundwater* is treated by six stages: pre-oxidation, GAC and slow sand filtration, aeration, neutralization and chlorine dioxide treatment.

The *lake water* treatment is as follows:

(a) pre-oxidation: with 0.3 mg/l Cl_2 and 0.6 mg/l ClO_2;
(b) micro-flocculation: with 1.5 mg^2/l aluminium sulphate;
(c) neutralization: raise pH 7.6–7.8 to pH 8.1–8.2. In Lengg with calcium hydroxide 6 g/m^3 and in Moos with caustic soda solution;

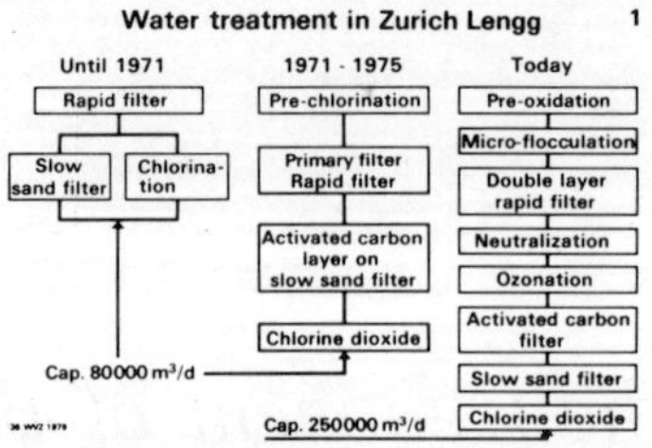

Figure 1

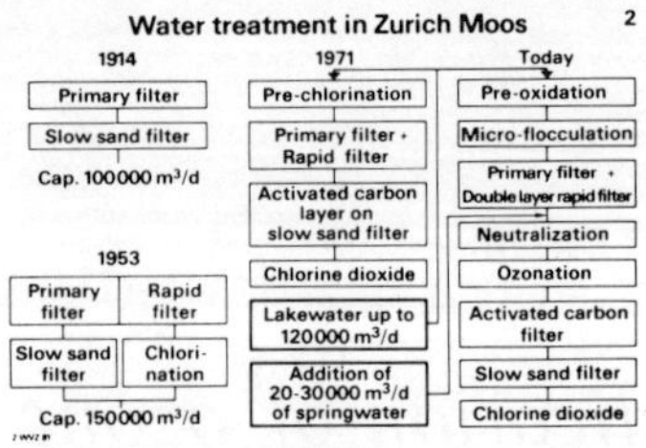

Figure 2

Comparison between the plants of 3

the plants of	Lengg	Moos
Plant capacity	250 000 m^3/d	150 000 m^3/d
Ozone production max.	32,1 kg/h	9,4 kg/h
Ozone addition normal	1,5 g/m^3	1,0 g/m^3
" " max.	3,1 g/m^3	1,5 g/m^3
" " max. (case of emergency)	14,2 g/m^3	7,6 g/m^3
Voltage max.	7,2 kV	13 kV
Frequency	350 Hz	50 Hz

Figure 3

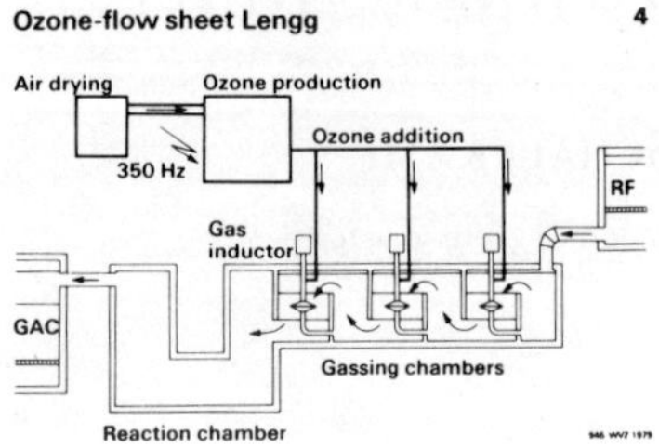

Figure 4

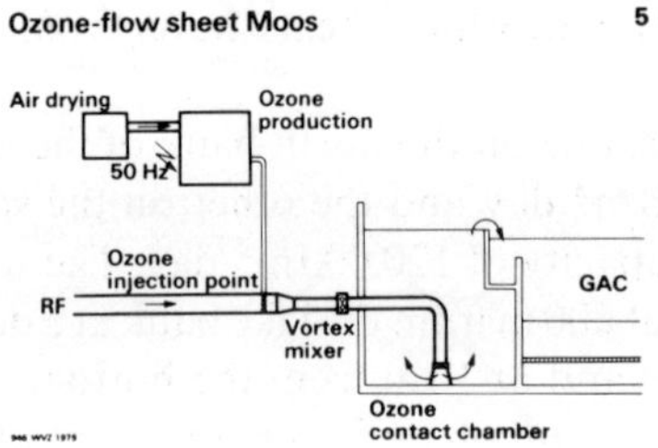

Figure 5

Technical comparisons of the ozone plants 6

		Lengg	Moos
O_3-production	Type	Tube-ozonator	Plate-ozonator
	Number of elements	6 à 1380 tube condensers	3 à 52 elements
	Frequency	Medium freq. 350 Hz	Normal freq. 50 Hz
		6 Freq.converters 50/350 Hz	No frequency converter
	Voltage	7,2 kV	13 kV
	Air treatment	Cooling and drying	Cooling and drying
O_3-mixing	Pressure conditions	Underpressure syst.	Pressure gas induction
		No compressors	3 compressors
	Gas induction	6 Radial injectors	5 gas induct candles with vortex mixer

Figure 6

Technical realizations of the ozone plants 7

		Lengg	Moos
O_3-Contact time at 40 % of plant capacity	Contact chambers	2 à 465 m^3 = 930 m^3	5 à 94 m^3 = 470 m^3
	cont.chambers	12 min	11 min
	GAC-overdam	14 min	19 min
	total	26 min	30 min
	System	Exhaust air from O_3-contact chambers GAC and retention pond Underpressure because exhaust air fan after catalyst	
O_3-exhaust air dissipation	Type of dissipation	Catalytic decomposition	
	Heating to	50°	110° goal 50°
	Catalytic mass	Filling 2×65kg	Filling 2×185kg
	Amount of exhaust air	2 × 650 m^3/h	170-510 m^3/h

Figure 7

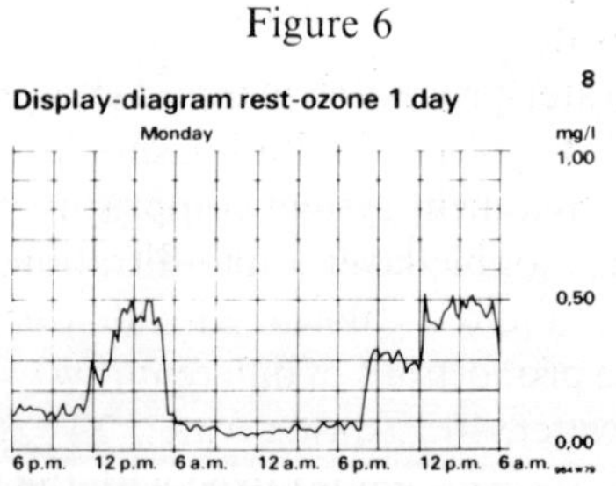

Figure 8

Effect of O_3 on various parameters 9

	1 mg/l	5 mg/l
Phenol	++	+++++
Viruses inactivation	+++++	+++++
Haloforms after 24 h influence	++	++++
Haloforms after 10 min. influence	+	++
UV-extinction 254 nm	++	+++
Colour	++	+++
KMn O4	+	++
Flocculations	+	-
Regermination	--	--
Aldehydes	--	----

+ positive; +++ very good; +++++ excellent; - negative; --- poor; ----- very poor

Figure 9

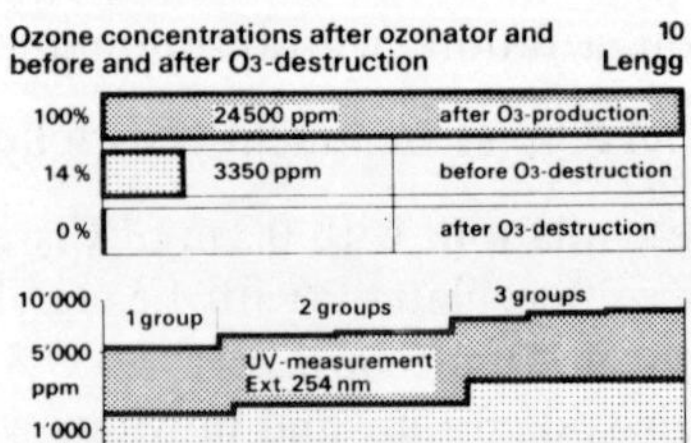

Figure 10

(d) rapid filtration: double-layer filter with 70 cm sand and 50 cm pumice, velocity: 6.0 m^3 per m^2/h;
(e) ozonization: all the information is contained in Figures 3–10;
(f) GAC filtration: double-layer bed with 50 cm sand and 1.2 m to 1.8 m GAC (in Moos Pittsburgh 400, in Lengg ROW 08 Super), max. velocity: 24 m^3 per m^2/h;
(g) slow sand filtration: river sand 50 cm to 100 cm, max. velocity: 15 m^3 per m^2/day;
(h) prevent-reinfection: 0.06 g/m^3 chlorine dioxide produced by reacting hydrochloric acid and an aqueous solution of sodium chlorite.

CHAPTER 75

Ozonization for groundwater protection at the City Waterworks of Karlsruhe

G. NAGEL
Stadtwerke Karlsruhe

One of the groundwater works of the City of Karlsruhe has been in operation for more than 110 years now and supplies $1400\,m^3/h$ or $6.5 \times 10^6\,m^3/a$. Within its catchment area a railway marshalling yard with dense transport traffic for oil products is located, from which the penetration of pollutants, mostly mineral oil products, cannot be avoided. The water is extracted from all levels of the aquifer, 25 m in depth, by four deep wells in line. With a permeability factor $k_F = 2.4 \times 10^{-3}$, the groundwater level is kept 4 to 5 m below surface.

Within the last years, the water from the nearest well showed an increasing contamination with organic substances, turbidity and UV-absorbing compounds. The oxygen and nitrate content of the groundwater there were also totally consumed. After abandoning the nearest well, the water quality in the neighbouring well deteriorated within a short period too. Figure 1 shows the location of the well line, the railway yard, five recharge wells, which are mentioned later, and the lines of equal groundwater levels.

At the same time cyanide-containing water, resulting from an abandoned chemical waste disposal site south of the well line, was also extracted from the ground. Therefore, the shutdown of the waterworks was programmed and could be foreseen.

To protect the precious watershed against a further spreading of the groundwater pollution, it was decided to use an ozone treatment for regeneration of the polluted aquifer. Preliminary laboratory tests had shown that ozone can oxidize the poorly degradable organic compounds in the groundwater, e.g. precursors and products of the crude oil industry, to improve and speed up biodegradation and mineralization. Within a research project, ozonization of the polluted well water was combined with a subsequent reinfiltration and underground passage.

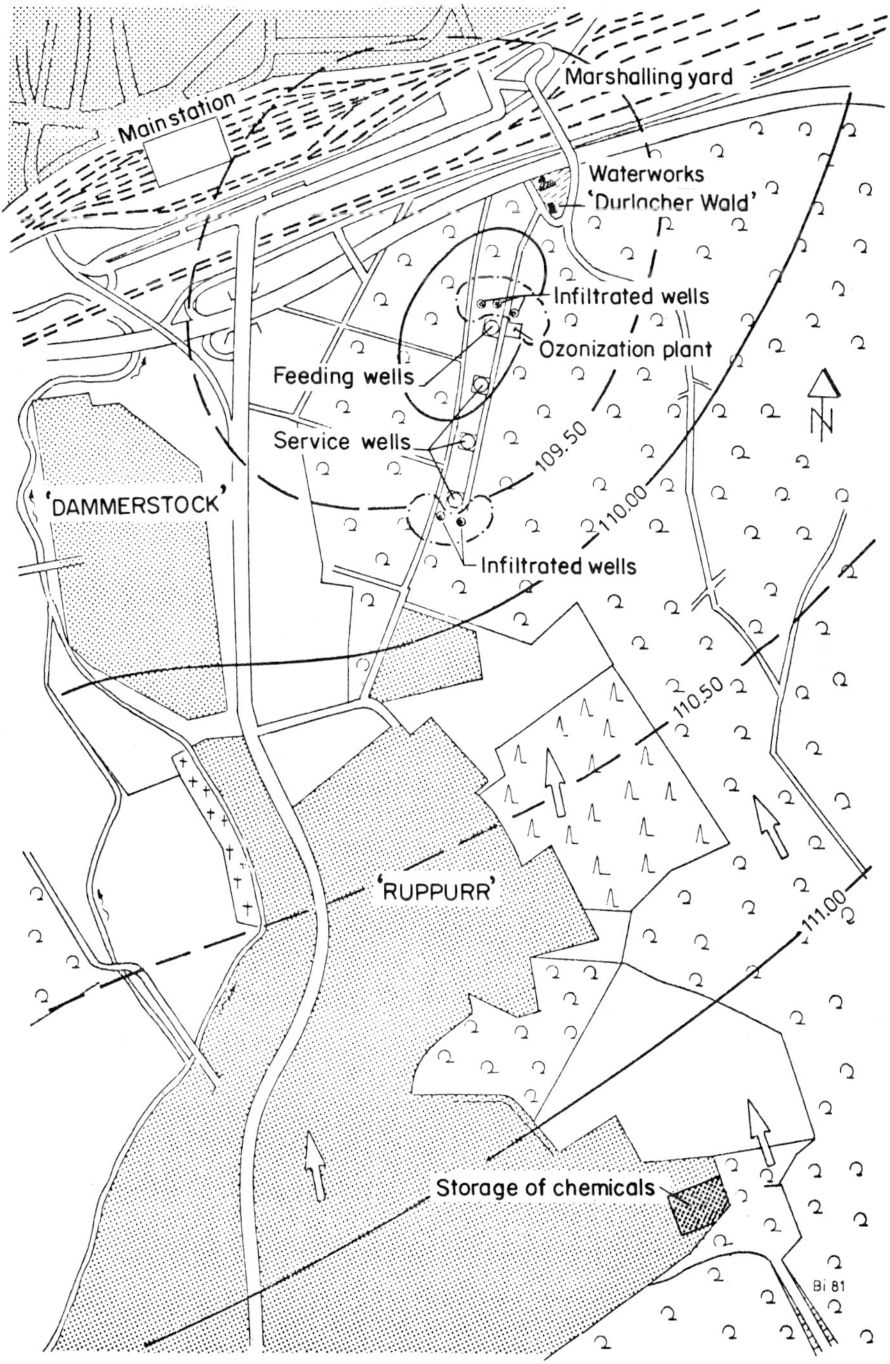

Figure 1. Location map of the waterworks 'Durlacher Wald', City Works of Karlsruhe, West Germany

Four infiltration wells were constructed between the marshalling yard and the first main well, 400 mm in diameter and penetrating the whole aquifer. Two wells of the same type were installed to the south, where cyanide-polluted water flows to the extraction wells.

Each of the infiltration wells is supplied with 100 m^3/h of ozone-treated water from the first main well with the highest contamination. In these wells, the water level rose between 0.5 and 0.6 m with a spreading radius of 60 to 80 m. With distances of 50 m between the reinfiltration wells, the water mounds overlap and a complete protection against the inflowing contamination is achieved.

Ozone is produced from dry air at a concentration of 25 g/m^3 and is introduced into the water by a static tube mixer and reacts with the organics in a bubble column (retention time 4 min) where the gas is separated at the top. The ozone in the off-gas is destroyed in a wet activated carbon filter. The water flows by the hydrostatic pressure to the infiltration wells.

Ozone is produced in a generator equipped with 120 tubes for electrical discharges at voltages of 15,000 V. The ambient air is dried to a dew point of below −45°C. The capacity of the generator is 1.2 kg O_3/h and can, if necessary, be increased to 2.6 kg O_3/h by adding pure oxygen. The ozone content of the carrier gas is kept at 20 to 25 g/m^3 since experience shows that under these conditions the best mass transfer and mixing efficiency can be achieved. The energy consumption for the ozone generation is 13.5 kWh/kg O_3, the additional equipment needs 6.5 kWh/kg O_3.

For mixing the ozone–air mixture (up to 50 m^3/h) with the water (up to 500 m^3/h) a static mixer is used, which is increasingly applied in drinking and wastewater treatment processes. Due to the lack of movable parts, it shows low wear and tear and needs hardly any maintenance. Tube mixers, with their intensive mass exchange, offer good mixing conditions with a low power demand.

Under laminar and turbulent current, the flow of products is stratified several times in the mixing elements of the static mixer. The subcurrents will become thinner with increasing stratifier number, thus raising the homogeneity. The number of stratifications can be described by a power law as a function of the amount of current channels in the mixing element and the number of elements in a mixer.

The power law was verified optically up to 100 subcurrents. Despite the better understanding of operation of static mixers by hydrodynamic laws, they are not suitable for a strict evaluation of the power needs. The success of a mixing process can be estimated quantitatively if the state of a mixture can be expressed by locally determined parameters. By a proper choice of tube diameter, type of mixing elements and element number, an economic static mixer can be designed for all mixing problems to achieve the projected mixing and intensive mass exchange.

The head loss in the mixer has to be compensated by the pumping and blowing equipment. At turbulent flow the resistance coefficient is constant, thus the

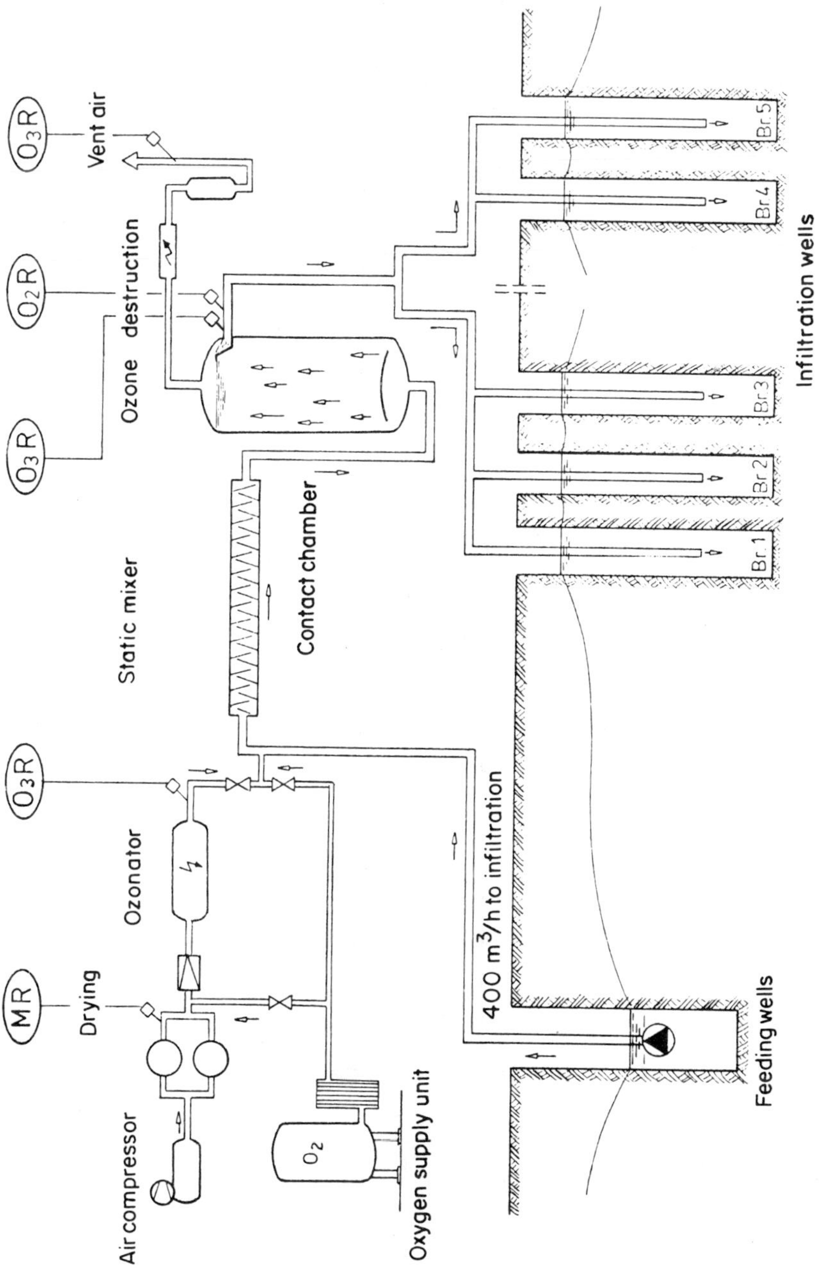

Figure 2. Operational scheme of ozonation plant and recharge facilities

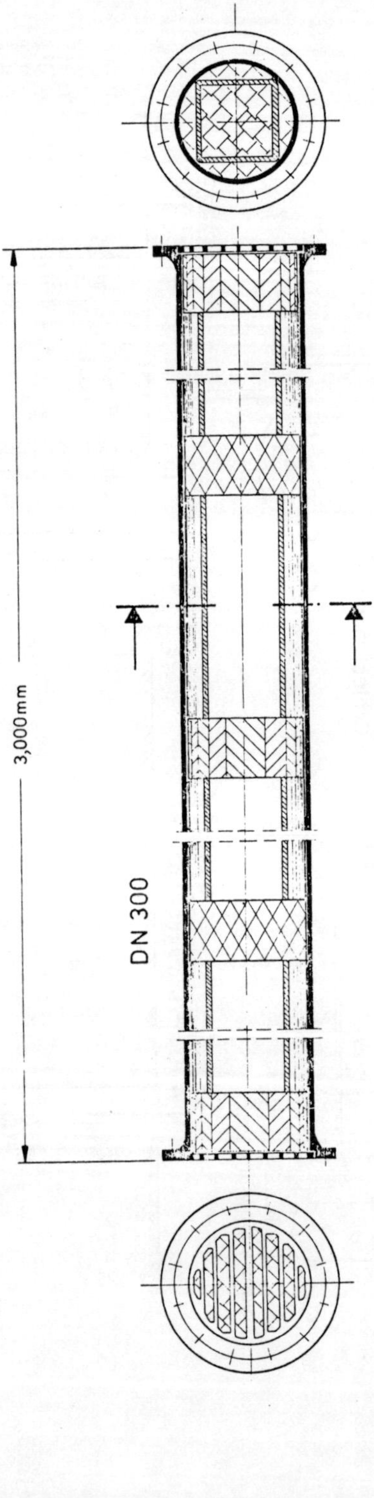

Figure 3. Static tube mixer

pressure loss is proportional to the square of the flow rate. For a certain geometry of mixing elements a constant Newton-number is gained. In contrast to this, the resistance coefficient at laminar flow is inversely proportional to the Reynolds-number.

In the case here, a tube, 300 mm in diameter, is used and the water flow rate of 1.36 m/s leads to a head loss of 2.1 m water column. With turbulent flow, a resistance coefficient of about 10 can be calculated.

A static mixer can be installed horizontally or vertically. In the latter case it can be operated in the up- and down-flow mode. The flow rate for in-line mixing of gases has to be higher than 0.7 m/s to achieve a sufficient mass exchange. By mixing gas and water at the volume ratio of 1:10, a mass transfer efficiency for ozone of up to 95 % can be achieved, determined from the ozone content of fresh and off-gas.

The recharged water is caught again in the main wells after 30 to 40 days in the aquifer. This contact time was sufficient to reduce the dissolved organic carbon (DOC) by biodegradation in the most heavily contaminated well within about 2 months after the start-up of ozonation from 3.5 mg/l to 1.5 mg/l. The regained water now contains several mg/l of oxygen. Bacteriological investigations indicated also an increased biological activity in the ground. The cyanide pollution was also controlled effectively by the recharge of ozonated water in the southern part of the catchment area. It can be projected now that the waterworks will remain in operation.

FURTHER READING

G. Nagel, Grundwasserschutz durch Beschickung von Abwehrbrunnen mit vorbehandeltem Wasser. In *Neue Technologien in der Trinkwasserversorgung*, Statusseminar des BMFT/DVGW, Aachen (1980), DVGW-Schriftenreihe Wasser No. 102 (1981)

H. Sontheimer and P. Wölfel, Amélioration de la dégradation biologique des eaux résiduaires par un traitement à l'ozone. *Int. Ozone Congress, Paris, May* (1977), IOA, Paris

H. Sontheimer, E. Heilker, M. R. Jekel, H. Nolte and F. H. Vollmer, *The Mülheim process. J.A.W.W.A.*, **70** (7), 393–6 (1978)

F. Grosz-Röll, Homogenitätserfassung in statischen Mischern. *Aufbereitungstechnik*, **20** (1), 26–30 (1979)

CHAPTER 76

The Dohne treatment plant in Mülheim–Ruhr: 'The Mülheim process'

M. R. Jekel
DVGW-Forschungsstelle am Engler–Bunte Institüt

Up to 1977 the Ruhr River water was treated at Mülheim by the classical combination of breakpoint chlorination, flocculation and sedimentation, rapid filtration, active carbon filters for dechlorination and infiltration into the ground for organics removal and storage. The concern about by-products formation in chlorination led to major changes in the treatment scheme, based on the results of an extended pilot plant study. The presently applied treatment sequence, called the 'Mülheim process' is depicted in Figure 1.

The Ruhr River water is first pumped to the rapid mixing basin (1.8 × 1.8 × 5.2 m) where the off-gases from the ozonization compartments are drawn in by a high-speed aspirating, turbine and are mixed simultaneously with the flocculant polyaluminium chloride. Additional fresh ozone from the generation plant can be added directly, to dose an average 1 g/m^3 O_3 within the 3 s retention time. Floc formation and separation is accomplished in the pulsator type of flocculator. The clarified water with less than 1 FTU is ozonized in basins (retention time 8 min, depth 4 m) where the gases are introduced by porous tubes at the bottom. A residual concentration of 0.3 g/m^3 of free ozone is kept in the effluent to ensure disinfection (total dose about 2 g/m^3). The ozonized water passes dual media filters and is then treated in the biologically effective activated carbon filters to mineralize biodegradable compounds formed by ozonization and nitrify ammonia (empty bed contact time 12 min). At times with high ammonia content, pure oxygen gas has to be dosed ahead of the rapid filters to ensure aerobic conditions at all stages.

The finished water is infiltrated into the ground, mainly for storage, and repumped after short retention times of 12–50 h, disinfected by 0.2–0.3 mg/l of chlorine and distributed.

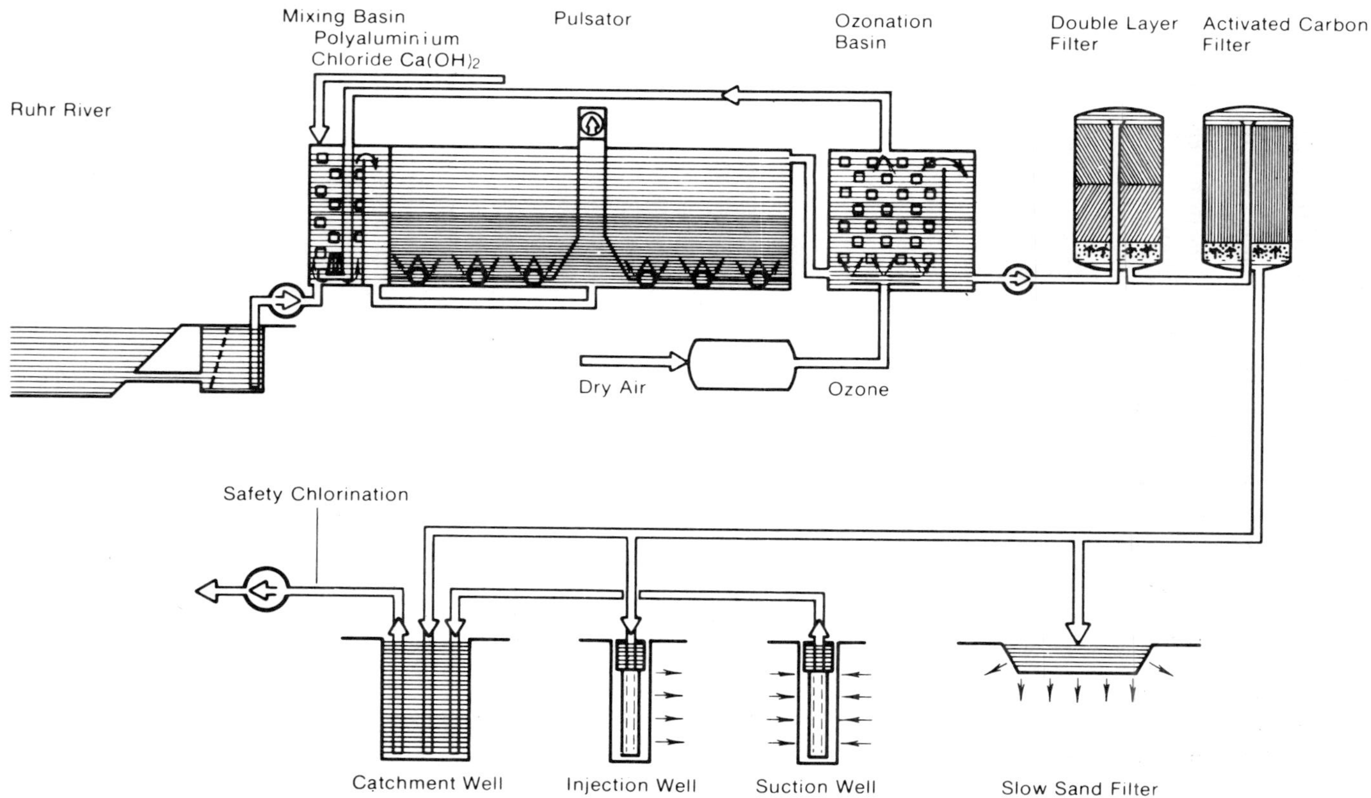

Figure 1. Process scheme for surface water treatment at the Dohne Water Utility in Mülheim after the introduction of combined chemical and biological oxidation

FURTHER READING

H. Sontheimer, E. Heilker, M. R. Jekel, H. Nolte and F. H. Vollmer, The Mülheim process, *J.A.W.W.A.*, **70,** 393 (1978)

E. Heilker, The Mülheim process for treating Ruhr River water, *J.A.W.W.A.*, **71,** 623 (1979)

M. Jekel and H. Sontheimer, *Biologische-adsorptive Trinkwasseraufbereitung in Aktivkohlefiltern*, Publications of the Engler-Bunte-Institute, Dept. of Water Chemistry, vol. 11. ZfGW-Verlag, Frankfurt/Main (1979)

CHAPTER 77

Ozonization and flocculation–filtration in the biological treatment of drinking water (the Essen plant)

A. REICHERTER

Stadtwerke Essen AG

The municipal services of Essen actually treat about 60×10^6 m^3 drinking water per year. Up to now, the principles of the treatment were essentially based on sedimentation and infiltration, the raw water source being that of the River Ruhr.

By mid-1981 a plant for exhaustive pretreatment of the raw water will be put into service. The treatment scheme is based on the following phases: ozonization of the raw water, flocculation of the ozonized water and subsequent filtration.

The water, pretreated according to that sequence, is then infiltered through existing slow sand filtration, infiltration basins. The sequence involved in this method is designed by the 'Essen process' and is the first existing sequence of this type. Its development has been supported by the Western German Ministry of Research and Technology.

The whole concept can be characterized by the following data:

Capacity: 15,000 m^3/h or 300,000 m^3/day.
Ozone: four ozonizers installed with a nominal production of 10 kg/h each.
Maximal permanent production: 34.5 kg O_3/h when using air.
Air flow: 1680 Nm^3/h.
Energy consumption: ± 1 MW.
Flocculation: with aluminium salts assisted with polymers.
Filtration: three-layer filters with a total layer thickness of 1.5 m composed (from lower to upper) by sand, carbon and activated carbon.
Surface loading: 15 m/h with an average cycle of 48 h (variable between 30 and 100 h depending on the water quality).
Filter surface: 100 m^2 installed in the two halls of each five filters (i.e. $2 \times 5 \times 100$ m)
Slow sand filtration: about 272,000 m^2 surface.

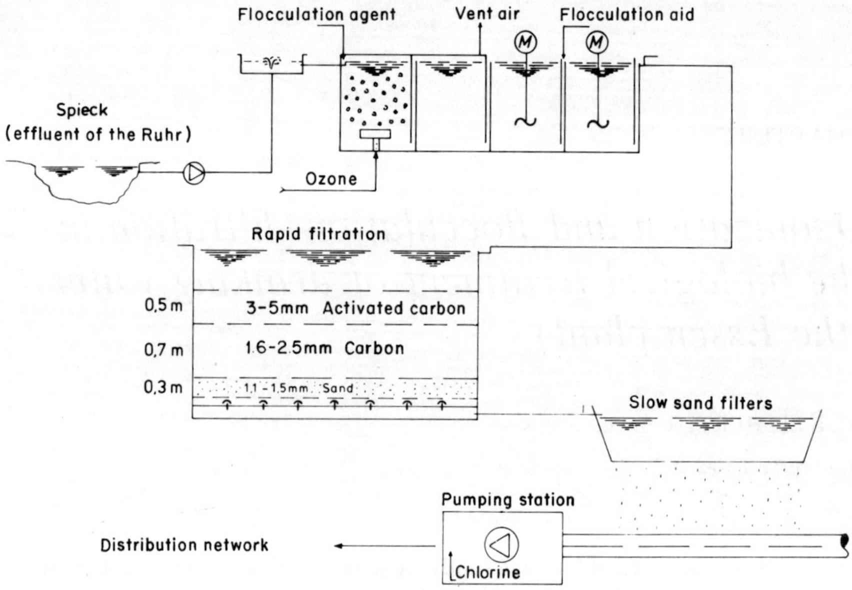

Figure 1. Flowsheet of the Essen plant

CHAPTER 78

The importance of ozonization in the purification system of the Lake Waterworks of Amsterdam

F. A. M. VAN SOEST

Gemeente Waterleidingen Amsterdam

The Lake Waterworks applies ozone in the purification process to have a hygienic barrier for bacteria and viruses, and to oxidize organic matter, in order to lower the taste, colour, UV absorption and DOC number.

The capacity of the plant is 30 million m^3/year. About 22 million m^3 is originated from the Bethunepolder. The 'polderwater', which has a low salt content of 40 mg Cl^-/l, contains quite a lot of humus-like substances. The rest of the required water will be drawn from the Amsterdam–Rhine canal, which is fed by the river Rhine. The mixture of polder water and Rhine water is the reason for the complicated purification system, which consists of the following sequences:

(a) separate coagulation of Rhine and Bethune water with $FeCl_3$;
(b) self-purification in the lake with a retention time about 100 days;
(c) rapid sand filtration;
(d) *ozonization* (dosage about 2.3 g O_3/m^3);
(e) if necessary, addition of powdered activated carbon;
(f) coagulation (dosage $FeCl_3$);
(g) rapid sand filtration;
(h) slow sand filtration;
(i) safety-chlorination.

The maximum and minimum production capacity of the ozonization installation (see Figure 1) amounts 25 to 2.8 kg/h.

The concentration of the ozone in the air varies between 15 and 25 g O_3/Nm^3. The ozonization as installed is performed in three consecutive chambers, where through porous tubes the ozone is introduced in to the water. The detention time in each basin is about 4 min. After the ozone-dosing chambers there are three contact chambers with a total detention time of 20 min (see Figure 2).

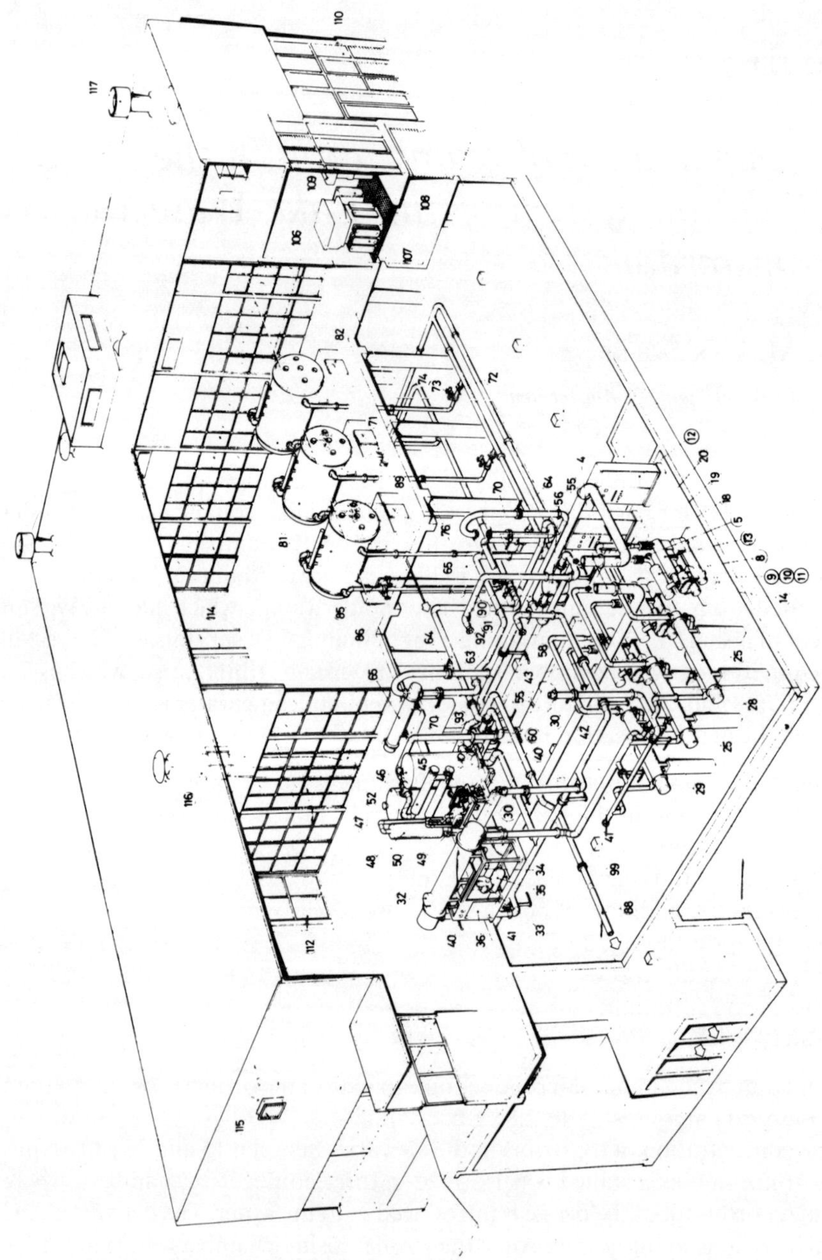

Figure 1. General view of the ozonization plant at Weesperkapsel (Amsterdam)

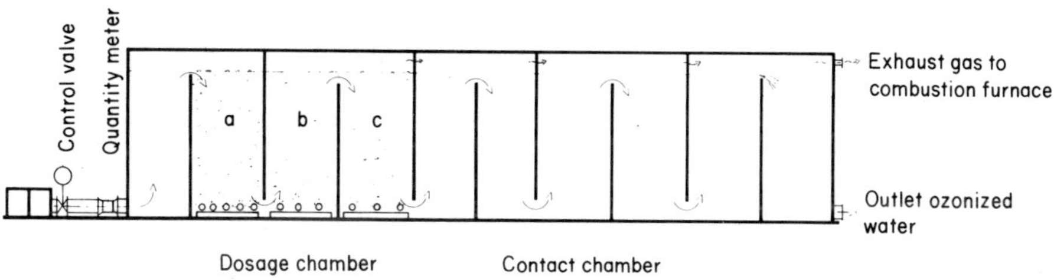

Figure 2. Cross-section of the ozonization chambers at Weesperkapsel (Amsterdam)

With only Bethunepolderwater the O_3 dosage is about $2.2\,g\,O_3/m^3$. The O_3 containing off-gas is destroyed by a catalytic ozone vent gas destroying installation.

FURTHER READING

De nieuwe Plassenwaterleiding van de Gemeente Amsterdam to Weesperkarspel, *H_2O*, **10**, 242 (1977)

CHAPTER 79

Ozonization of recycled water at Quierschied's swimming pool

J.-P. LEGERON

Trailigaz

The Quierschied swimming pool is covered and equipped with a wave generator. Its dimensions are as follows: 33.3 m long, 12.5 m wide and from 0.8 to 2.8 m deep.

The pool is supplied with treated water through eight inlets located on one of the longitudinal walls. They are placed at 0.8 m under the water surface and are at alternate intervals of 1.5 m and 3 m from the deeper end of the pool. Water recycling is purveyed both from the surface (30 % of the total flow) and from the opposite side of the pool (70 %) through eight outlets of similar shape as the inlets.

Water leaves the swimming pool at a 150 m^3/h flow, and is treated as presented as shown in Figure 1: prefiltration is the first treatment step for the water coming

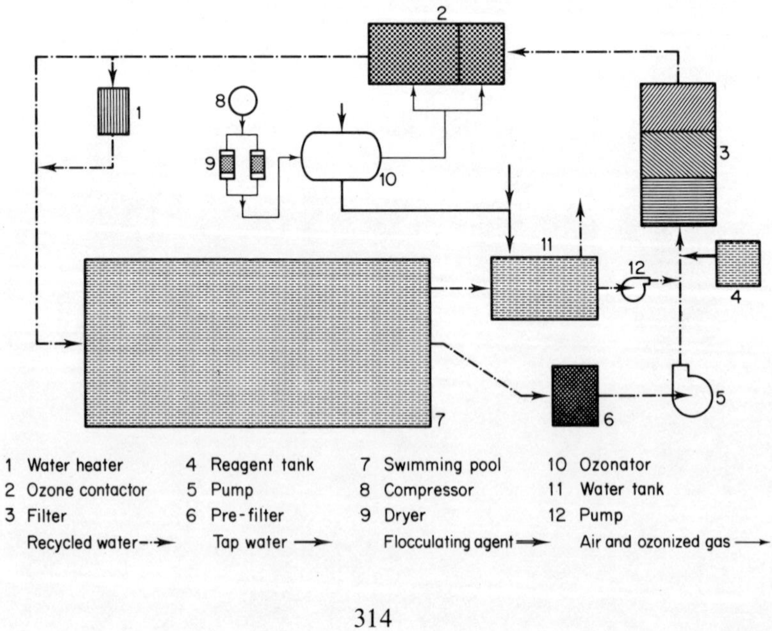

Figure 2. General view of the swimming pool

from the eight outlets. As for the overflow water, it goes into a tank where it is mixed with the daily make-up water; i.e., 29.6 m^3/day including the water necessary for cooling the ozone generator. After pumping, the waters from the two circuits are mixed and small amounts of aluminium sulphate are added. A three-layer filtration is applied comprising sand, anthracite and activated carbon. Subsequently ozonization is practised: this is made in a stainless steel pressurized contact column which is composed of three chambers: in the first, ozone and water circulate counter currently, so as to obtain a free dissolved ozone residual of 0.4 mg O_3/l; in the second, ozone and water circulate equicurrently and the 0.4 mg O_3/l is maintained during 4 min. In this ozonization scheme the conditions for bacteria and virus destruction are met. Finally, in the third chamber, a gas-stripping eliminates most of the dissolved ozone remaining in the water.

Before flowing back to the pool, water is warmed by deriving part of the water flow in a heating unit.

Ozone is generated by an Ozobloc type generator. The treatment rate is 0.5 g O_3/m^3 of water during the day, and 0.75 g O_3/m^3 during the night. In the contact column, ozone diffusion is achieved through porous ceramic diffusers placed at the bottom of the chambers. The transfer reaches 94 to 97% yield.

With such a treatment, a water of perfect quality is offered to swimmers without any further addition of other disinfecting agents to the water.

CHAPTER 80

Ozonization at the swimming pool of Ostend

W. J. MASSCHELEIN

Compagnie Intercommunale Bruxelloise des Eaux

The whole complex of the swimming pool of Ostend comprises two covered basins; the first, an Olympic pool of a capacity of 2.325 m^3 (see Figure 1), the second, a learning pool of a capacity is of 120 m^3, and two outside pools of respective capacities of 1,100 m^3 and 100 m^3. The feedwater is impounded from the North Sea and prefiltered on rapid sand filters of the open type. The recycled substream, common to all basins, is designed as 600 m^3/h.

However, the outdoor pools are of a lower level than the indoor ones and therefore, in order to avoid water transfer from one system to the other, there is a level regulator which controls the modulating valves of the return circuit of the outdoor pools.

In the recycling circuit the water is flocculated and filtered on pressurized sand filters of the horizontal type.

Facilities for ozonization of the treated water of the substream are installed by a horizontal tube ozonator of a nominal production capacity of 960 g/h; i.e. 1.6 g O_3/m^3 treated and recycled substream. The practical injection dose is adapted to maintain 0.4 g/m^3 as residual concentration for at least 3 to 4 min following the two-fold injection chamber. Excess ozone is stripped through a cascade at the inlet of the basins, and hydraulic waves can be induced in the main indoor pool.

Ozone contacting is obtained by injection through porous pipes into three stainless-steel pressurized contact chambers mounted in parallel to each other to treat 200 m^3/h of the substream flow. The injection and contacting towers are subdivided into two successive compartments enabling the immediate contacting phase and the ozone residual maintaining phase of 3 to 4 mins.

The water returned into the basins is perfectly disinfected and pleasant to the swimmers, although no significant ozone concentration is maintained in the basins, an active residual oxidant concentration is revealed through the usual tests, e.g. O-toluidine or DPD. An active residual such as 'bromine' is likely to be formed on ozonization of the bromide in the sea water. This soft disinfectant is less aggressive to the mucous membranes and allows an ideal combination ozonization of water containing bromide, e.g. sea water, or bromide-enriched recycled water.

Figure 1. General view of the Ostend's indoor Olympic sea water pool

Index

Absorption (physical), 47
Absorption (UV), 191
Acceleration (factor), 50
Accuracy, 163, 175
Acid chrome violet K, 164, 166
Acid (malonic), 170, 171
Acids (carboxylic), 227
Actinomycetes, 94
Activated carbon, 64, 144, 212, 214, 217, 218, 259
ACVK, 164, 166, 167
Adsorption, 34, 221
Aeration–filtration, 286
Aftergrowth, 100, 224, 227
Air (drying), 146
Alcohols, 76, 103
Aldehydes, 61
Algae, 69, 88, 288
Alginate, 271
Aluminate of calcium, 142
Ames (test), 198
Amines, 73, 74, 198
Aminochloracetic acid, 163
Ammonia, 80, 224, 243, 271, 283
Ammonium chloride 182
Amperometry, 176
Amsterdam, 311
Annet-sur-Marne, 286
Arnolds reagent, 164
Aromatic (compounds), 83
ASTM-standard, 160
ATP, 84, 248
Augers, 288
Auto-decomposition, 99
Automatic control, 37
Automation, 177, 178, 179

Backflow (of air), 147
Bacterial competition, 228
Bacterial indicators, 226
Benzopyrene, 256
Biodegradability, 64, 83, 253
Biodegradation, 84, 305
Biological (filtration), 243
Biomass, 84, 248
Bleaching, 258
BOD, 248, 250, 257
Borate buffer, 160
Bottled water, 259
Breakpoint, 184
Bromate, 80
Bromide, 61, 79
Bromine, 317
Brussels (plant), 269
Bubble-dispersing, 137
Bubble radius, 138
Bubbles, 94, 100, 188
Bubbles (gas), 93
Bubbles, rigid and soft, 138
By-products (of ozonization), 105

Calorimetry, 155
Capacitive current, 188
Captane, 198
Carbon (activated), 82, 83
Carbon dioxide, 63
Carbon tetrachloride, 198
Catalysis, 204, 207, 212
Catalytic (disposal), 234
Catalytical decomposition, 214
Ceramics, 142
Chemi-luminescence, 155, 159, 164
Chemionuclear synthesis, 23
Chloramines, 163, 181, 183
Chlorate, 90–91, 173
Chlorination (post-), 224
Chlorine, 144, 183, 227
Chlorine dioxide, 90, 91, 92, 183, 225, 269
Chlorine (interference), 163
Chlorite, 173, 225

Chloroform, 198
Choisy-le-Roi, 280
Chromosomic aberrations, 201
Clarification, 81
Cleaning (electrodes), 188
Coagulation, 86
Coating (metal), 33
Coating systems, 145
Coatings, 143–144
Cobalt, 173
COD, 250, 257
Colloids, 88
Colloids (protected), 88, 93
Colony counts, 259
Colour, 63, 250, 256, 257, 278
Combustion, 222
Competition values, 59
Compression (of ozone), 21
Computer, 180
Concrete, 142, 244
Constance, Lake, 267
Consumption (of ozone), 212
Contact columns, 137
Contacting, 234
Contacting (of ozone), 47, 129, 317
Contacting system, 129, 176
Contact time, 133, 136, 305
Continuous analysis, 164
Controls, 27
Converter (frequency), 35
Cooling, 26, 42, 146
Co-oxidation, 228
Copper (electrode), 181
Copper–gold electrodes, 176
Corrosion, 140, 143, 188
Coryneiform bacteria, 227
Costs, 76, 101, 104, 138, 178, 179, 209, 215, 216, 217, 219, 220, 221, 233, 235, 238, 239, 243, 257, 254
Counter-current exchange, 130
Counter-current flow, 244
Crissey (plant), 271
Cyanides, 103, 250, 253, 300, 302

Damage, 146
Decoloration, 86
Decomposition, 26
Decomposition (of ozone), 53, 59, 209
Degasification, 18
Demicellization, 93
Density (current), 38, 40
Depolarization (current), 188
Design (cost), 238
Destruction (of ozone), 5, 217
Destructor, 209
Detection (of ozone), 161
Detention time, 134, 243
Determination (of ozone), 151
Dew point, 34, 147
Dexon, 198
Dielectric, 28, 33, 36, 38, 177
Diffusion, 144, 147
Diffusion (coefficient, 49
Diffusion (of ozone), 93
Diphenylhydrazine, 198
Discharge, 4, 6, 33
Discharge (electrical), 7, 9, 13, 40
Disinfection, 66, 178
Disposal (ozone), 234
Dissolution, 18, 93, 130
Distribution system, 225
Dose (of ozone), 83
Double beam photometer, 189
Double film, 47
DPD, 163
Drying (air), 179, 234
Dust, 25
Dyes, 254

Earth, 38, 40
Earthing, 33
Economics, 17
Efficiency, 178
Efficient (sections), 7
Effluents (industrial), 253
Electrical field, 5, 7, 11
Electrical power consumption, 35
Electrode conditioning, 188
Electrolyte (analysers), 161
Electron impact, 9
Electrophilic (reactions), 74
Energy (activation), 4
Energy consumption, 27
Energy (for destruction), 213
Energy (internal), 9
Energy recovery, 207
Enthalpy, 3, 153, 159
Epoxidation, 74
Epoxides, 197
Errors, 158
Essen, 309
Excess ozone, 204

Excitation states, 4
Exhaust gas, 18
Explodability, 21
Explosivity, 221
Exposure, 200, 202
Exposure (to ozone), 140

Fenitrothion, 75
Flexibility, 82
Flocculants, 64
Flocculation, 63, 87, 88
Flocs, 100
Flotation, 249
Flowrates (in catalytic contacting), 208
Foaming, 142
Foam (in ozonization), 249
Formation of ozone, 3, 7
Fouling, 188
Frequency (current), 30, 40
Furnace (heat), 206
Fuse-system, 31, 36

Galvanic current, 176, 181
Gas (composition), 11
Gas-masks, 202
Gas-transfer, 137
Generator, 33, 24
Glycine, 163, 170, 171
Glyoxal, 198
Glyoxalic acid, 198
Gold (electrode), 181

Half-life time, 260
Haloform reaction, 77
Harmonic currents, 31
Harmonics, 34
Hartley-bands, 153, 157
Heat, 204
Heat (decomposition of ozone), 210
Heat effect, 3
Heat exchange, 209
Heat exchanger, 220
Heating, 218
Henry's law, 48
Heterogenicity (of bacteria), 229
Hospital-use, 199
Humic acids, 64, 89, 93, 198
Humic compounds, 63, 88
Hydraulic circuit, 244
Hydrocarbons, 255, 256
Hydrogen peroxide, 173, 200
Hydroperoxides, 174
Hydroxyl ion (action on ozone), 54
Hypalon, 142, 202

Inactivation (of bacteria), 98
Indigo derivatives, 169
Indigo trisulphonate, 169, 170
Intensity (current), 31
Interferences, 166
Intermediates, 4, 7
Investments, 16
Iodide (potassium), 152
Iodide (zinc), 152
Iodometry, 151, 154, 157
Iron, 82, 86, 88, 93, 198, 249, 259, 271, 279, 288

Karlsruhe, 300
Kinetics, 4, 129, 134
Kinetics (of ozonization), 95
Kinetics of reaction, 131

Lake water, 63, 69, 86, 90, 311
Lengg (plant), 297
Leucomalachite, 164
Limiting nutrient, 228
Losses (of ozone), 131, 132, 220

Maintenance, 42
Malathion, 75
Maleic acid, 198
Malonaldehyde, 199
Manganese, 64, 82, 86, 88, 93, 134, 166, 169, 198, 259, 271, 283, 288, 291
Manley (relation), 28
Marseille, 293
Mass transfer, 95, 100, 129
Measurements (continuous), 157
Mechanisms (reaction), 3
Méry-sur-Oise, 280
Mesophilic bacteria, 229
Metabolization, 227
Micellization, 86, 88, 243
Micrococci, 227
Microprocessor (control), 180
Mineral water, 259
Mixing, 94, 304
Mixing (static), 302
Modelling, 6
Mohlman index, 249
Moisture, 25

Monod's equation, 228
Monitoring, 147, 157, 181
Monitoring (UV), 158
Moos (plant), 297
Morsang-sur-Seine, 276
Mucosa, 197
Muenke flasks, 151
Mülheim/Ruhr, 306
Multiple ozonization, 81
Mutagenicity, 201

Neuilly-sur-Marne, 280

Nice, 265
Nickel–silver electrodes, 187
Nitrates, 151
Nitric acid, 10, 24, 142, 146
Nitrification, 243, 283
Nitrifying bacteria, 271
Nitriloacetic acid, 198
Nitrites, 250
Nitrogen bacteria, 224
Nitrogen oxides, 10, 11, 24, 151, 159, 200, 208, 217
Nucleic acids, 67
Nutrient (limiting), 228

Odour, 257
Off-gas, 217, 220
Oil, 146
Oil (refining), 255
Ondulator, 30
Operation cost, 239
Optimization (of ozone concentration), 235
Organic compounds, 73, 87
Organo-chlorinated compounds, 77
Organophosphorus compounds, 75
Ostend, 317
Overheating, 147
Oxalic acid, 198
Oxidations (induced), 61
Oxygen, 13, 18, 38, 305
Oxygen (process gas), 238
Oxygenation, 243
Ozone concentration, 23
Ozone contact, 185
Ozone dose, 83
Ozone losses, 131
Ozone removal, 212
Ozonide, 166

Packed columns, 137
Palladium, 213, 218
Papermills, 257
Paris, 276, 280, 286
Particle size, 88
Perborate, 174
Percarbonate, 174
Perchlorate, 92
Perlauric acid, 166, 174,
Permanent operation, 237
Peroxide, 99, 103, 169, 197
Peroxydic compounds, 74
Pesticides, 73, 74
Phase exchange, 129
pH-correction, 188
Phenols, 58, 78, 198, 250, 251, 255
Photochemistry, 22, 103
Photolysis, 221, 222
Photometric determination, 153
Photometry, 155, 191
Picric acid, 94
Pilot plant, 38
Plankton, 291
Plasma's 3
Plastics, 260
Plate columns, 137
Platinum, 218
Platinum–copper electrodes, 176
Polarity (molecular), 63
Poliomyelitis, 67
Polyamides, 74
Polymerization, 86
Porous diffusors, 138
Post-ammoniation, 269
Post-disinfection, 64
Post-ozonization, 238, 285
Post-treatment, 225
Potentiometry, 178
Power consumption, 37
Power (electrical), 40
Power input, 28
Power requirements, 205
Precision, 175
Precursors, 106
Precursors (THM), 77
Preoxidation, 286
Preozonization, 69, 283, 288, 306, 309
Pressure, 25, 40
Pressure drop, 207
Process gas, 16
Production, 39, 42

Production capacity, 16, 24
Production cost, 234
Production (ozone), 11, 13, 36
Professional risks, 200
Protection, 160
Protection (personal), 202
Pseudomonas, 227
Pulmonary oedema, 197
Pulp (mechanical), 258
PVC, 142

Quantum Yield, 222
Quierschied, 314

Radiations, 11
Radicals, 6, 9, 22, 103
Radicals (OH′), 59, 61
Reaction (pathway), 57
Reaction products, 105
Reaction (rates), 57
Reaction time, 104
Recovery (of ozone), 235
Recycling, 16, 18, 20, 238
Recycling (of water), 243, 314, 317
Redox potential, 144, 260
Redressor, 31
Reduction, 23
Regeneration, 147
Regulating, 176
Regulation, 289
Regulator (inductive), 29
Reinfection, 64
Relay protection, 32
Rennes, 289
Residual ozone, 131, 162, 166, 173, 179, 181, 243, 317
Resistance (corrosion), 21
Resorcinol, 78
Respiratory tract, 197
Reynolds number, 138
Rhodamine B, 96
Roberval (Canada) plant, 278
Ronnel, 75
Rouen-la-Chapelle (plant), 283
Rubber, 142, 143
Rubber lining, 145

Safety, 146
Salmonella, 198, 249
Sand (filtration), 70, 71
Saturation, 129
Saturation (in ozone), 129
Scrubbing, 223
Sea-water, 318
Secondary effluent, 249, 251
Secondary treatment, 248
Security, 203
Security (electrical), 31
Semi-conductor, 30
Short-circuiting, 36
Sipplingen (plant), 268
Solubility (of ozone), 53
Sonication, 98
Spark discharge, 10
Spark phenomenon, 36
Spatial velocity, 217
Specifications, 24
Specific heat, 205
Specificity, 163
Spray-columns, 137
Stability (of ozone), 53
Stainless steel, 210, 244
Stationary counts, 228
Steady state (inactivation of bacteria), 98
Storage, 81
Storage (oxygen), 14
Substream injection, 137
Sulphur dioxide, 200
Sulphuric acid, 23
Surfactants, 250
Suspended matter, 248, 250
Swimming pool, 243, 314, 317
Synergism, 11, 200
Synergistic action, 96
Synthesis, 73
Syringaldazine, 163

Tailfer (plant), 220, 239, 269
Tele-control, 296
Telluric bacteria, 227
Temperature, 26, 143, 182
Temperature effects, 66
Tension (electrical), 8
Tertiary effluent, 251
Tertiary treatment, 249
Textiles, 254
Thermal decomposition, 205, 234
Thermal destruction, 212, 214, 217
Thermodynamics, 6
THM, 103
Thyristor, 30, 31, 37
TOC, 250, 257

Toluene (dinitro-), 198
Toxical products, 73
Toxicity, 197, 200, 204, 248
Transfer (of ozone), 47, 49, 93, 129
Transformer, 34, 38
Transformer (auto-), 30
Transformer (Scott), 29
Trihalomethane, 78, 80
Trihalomethane (formation potential), 77, 78
Turbidity, 88, 250
Turbines, 138, 306

Ultrasound, 96
Ultraviolet (UV) radiation, 103
Uprating of ozone production, 238
UV absorption, 83, 151, 157, 176, 185, 189
UV decomposition (of ozone), 222
UV irradiation, 22
UV lamps, 223

Valence, 291
Velocity gradient, 94, 138
Viruses, 66, 82
Voltage, 28, 177

Water vapour, 5
Wastewater, 199, 248, 253, 257
Wastewater treatment, 212
Work areas, 160

Yields, 3, 6

Zeolithes, 221
Zeta potential, 249
Zinc chloride, 152
Zooglea, 85
Zurich, 297